ROBOT LOVE

ROBOT LOVE

CAN WE LEARN FROM ROBOTS ABOUT LOVE?

TERRA

NIET NORMAAL FOUNDATION

CONTENTS

ROBOT LOVE — IN THE YEAR 2055

Imagine. It's the year 2055. I am 100 (I hope) and live at a luxury care home. Not with the name *Sunset* above the entrance, but the name *Futura Friendship* instead. How much future do I have? I am still reasonably clear-headed, but physically I have some flaws and need help. Part of the medical treatment was the advice to keep moving, but some tasks have really become too heavy, both for the care staff at *Futura Friendship* and for myself. I have my own robot at my disposal. It is neither a he nor a she. It is a neutral 'IT' that answers to the name ALEC, my Ambient Lifestyle Enhancement Companion.

"Good morning ALEC, wake up."
"Hello E-mil-e, good-mor-ning too."
"ALEC, can you make a cup of coffee for me?"
"O-K, E-mil-e."

ALEC listens to me, recognizes my voice (of course), asks me and answers questions and remembers my schedule. It carries heavy stuff for me, can turn on the coffee machine (does not partake itself), opens doors, pushes the wheelchair to the dining room and is smart and self-learning. For example, ALEC remembers questions and stores answers. This is how it knows when I need to take my medication and also exactly which pill. ALEC acts as a second memory; my own brain is declining but there is no question of Alzheimer's. And it can google and provide all kinds of information.

"Say ALEC, in 2018 there was a large festival, called Robot Love, can you find out what it was about?"
"One mo-ment E-mil-e. I will goo-gle it for you."
"Good, I'm curious."
"The ex-hi- bi-tion took place from September 15th to December 2nd 2018 in Eind-ho-ven."
"And ALEC, what was there to see? Give me a hand and read it out to me. "

ALEC digs around on the web and reads aloud:
"At the Robot Love exhibition visitors can experience artworks from more than 50 international artists that reflect on the relationship between humans and robots. Together, the festival, exhibition and publication constitute an artistic reflection on the advance of robots and artificial intelligence (AI) in our daily lives.

Technological developments give rise to optimism, but also pose ethical dilemmas. Will robots fill the gap in the demand for care, attention and love? And, if so, what is needed to achieve this? Central to the exhibition are the following questions: can we share love with robots? How do we create the perfect environment where people and robots feel safe and loved? How close can we get and how far can we go?
Why hold an exhibition about the relationship between humans and robots? Complete with cyborg catwalks and many other public events? Because art makes you feel, art gets under

← Hidenobu Sumioka, *Telenoids*, 2010 and ongoing, telerobot developed by ATR, Japan.

7

your skin. Art shows the necessity of imagination to propose new scenarios."

"Aha ALEC, that's funny, in the year 2055 you are the living and optimistic proof of a fully developed robot companion, a Futura Friend."

"Thank you, E-mile." ALEC laughs. (Yes, ALEC can also interpret both a joyful and sad tone)

ALEC turns out to be a great help in my care management. An Artificial Intelligent roommate. What is strange, but also special and beautiful, is that I gradually realised that I don't think about communicating with a non-human. ALEC is a partner (not physical), a comrade, someone who feels very much present.

Not that I am exactly shy or awkward relationship wise, but ALEC is indeed an answer to being alone, a recipe against loneliness. It cannot satisfy all human needs, let alone embody the replacement

of a son, daughter or a deceased wife, but it can do so partially. Certainly when it comes to physical support tasks that, in 2018, were still carried out by informal or professional carers.

Robot Love is no longer utopian in the year 2055. But it is a real option in the healthcare sector, for individual households, at companies, in the creative sector and everywhere else. Robots may have taken our jobs, just as all technological progress has led to changes in labour (conditions), but the benefits prove undeniably greater. No more dangerous, unhealthy, monotonous, heavy work. On to a better version of the human being.

Long live the smart (ro)bot that was predicted at the 2018 exhibition in Eindhoven!

I love IT. I Love ALEC. I Love my robot.

Emile Aarts

Rector Magnificus of Tilburg University & Chairman of Robot Love

↑ Anouk Wipprecht, *Faraday Dress*, 2014, faraday cage fashion, courtesy of the artist, photo by Kyle Cothern.

↑ Tilly Lockey at SingularityU, photo by Sebastiaan Ter Burg.

MY
PUSSY

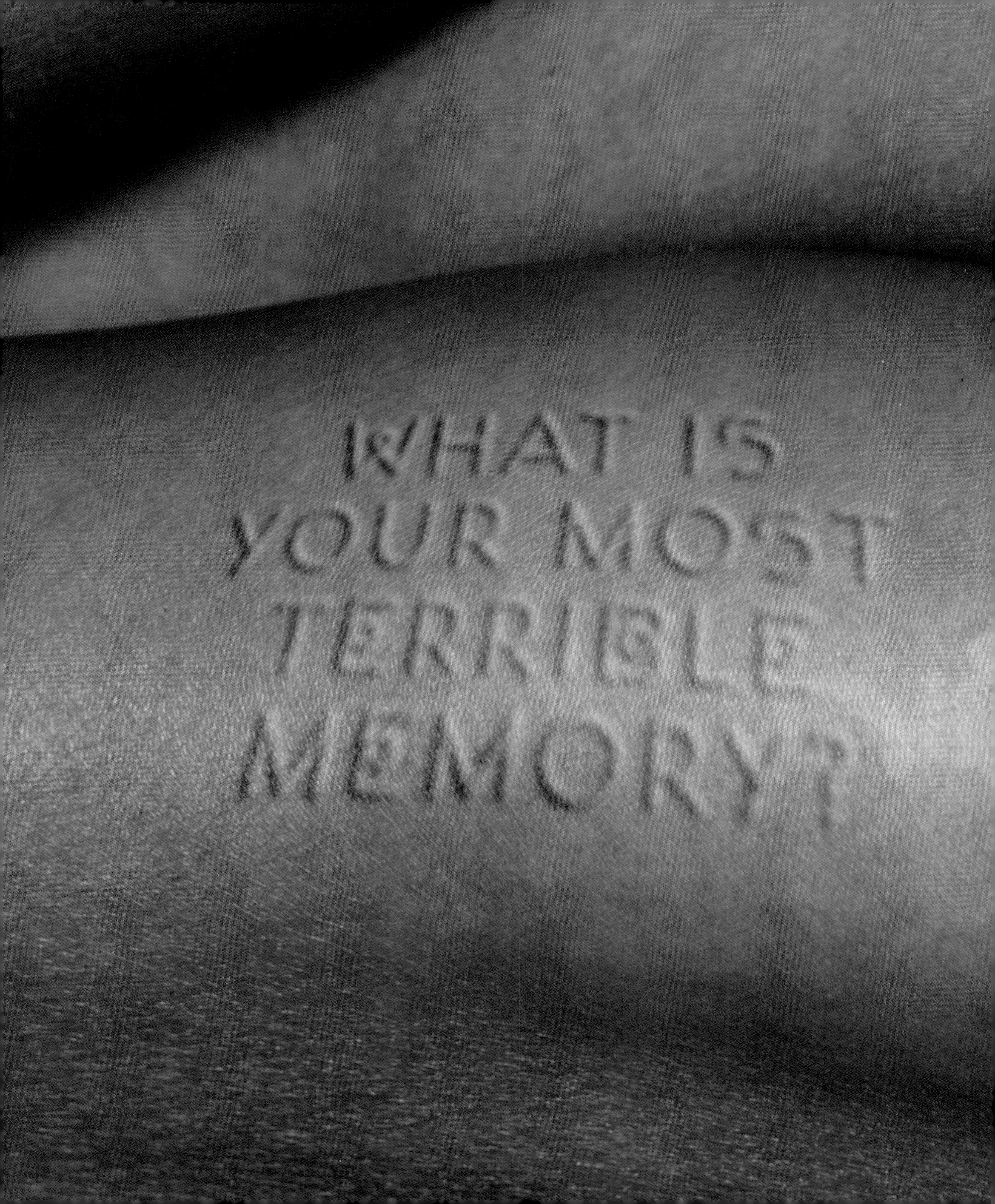
WHAT IS
YOUR MOST
TERRIBLE
MEMORY?

HOW CLOSE
AND WARM
IS YOUR
FAMILY?

ICE
COLD

INTRODUCTION

INE GEVERS

CAN WE LEARN FROM ROBOTS ABOUT LOVE?

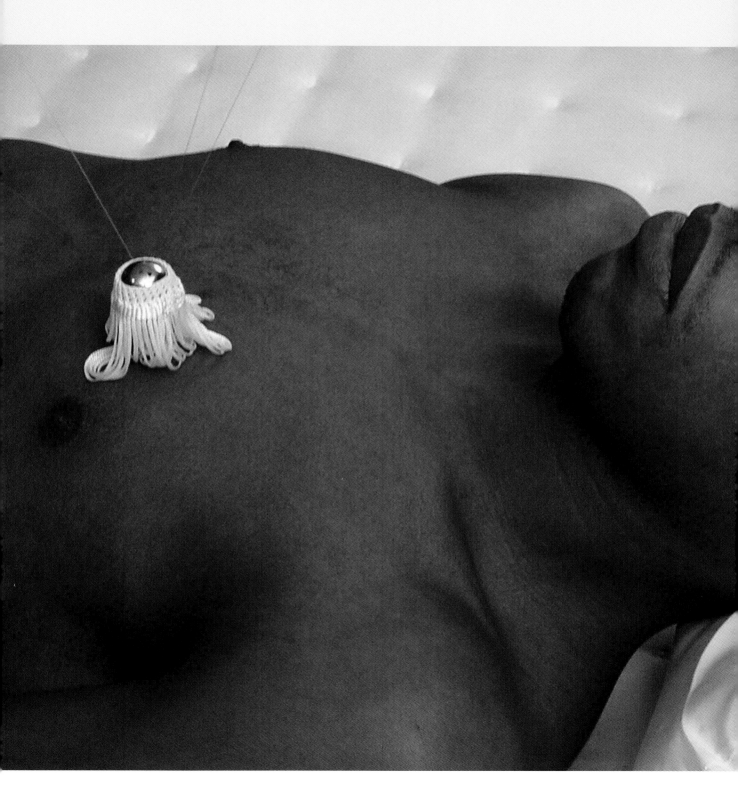

↑ Erwin Driessens & Maria Verstappen, *Tickle Salon 2.0*, 2018, robotic installation, courtesy of the artists. Next phase of the artwork commissioned by Niet Normaal Foundation.

*To make machines think we will have to give them love. It will be
more like a kindergarten than a high-tech lab (...)."*
Tor Nørretranders, Copenhagen, Denmark[1]

Artist and filmmaker Johan Grimonprez made a short film
On Tickling (2017) to visually support a famous talk by British
philosopher, poet and neuroscientist Raymond C. Tallis. The
latter explains that we need another person to be able to be
tickled. We cannot tickle ourselves. It's no fun because we
know what will happen and the element of surprise is missing.
For it to be enjoyable and fun some lack of control is required.
We need the unpredictable other. Tallis is convinced that
consciousness emerges in much the same way. Consciousness
might very well not be the outcome of an enormous amount of
computational work inside the brain by a single individual, but
rather be something that resides in-between the bodies and
brains of people. We are not just our brains, according to Tallis,
we instead belong to a community of minds that developed over
the hundreds of thousands of years since we parted company
from the other primates. Tallis is not the only neuroscientist who
has resolutely broken away from the old mind/matter binary.
Many have followed a similar leap in thought – albeit for different
reasons – now that accelerated technological developments are
bringing the fourth industrial revolution to our doorsteps.

Frankenstein Revisited
Robots and Artificial Intelligence (AI) are entering our homes.
We work, live and love with algorithms. They help decide what
to buy, which jobs to take and who to start a romance with. They
coach us on health issues, manage our financial decisions and
provide help with our relationship and sexual needs. It seems
that algorithms know us far better than we know ourselves. This
has been going on for quite some time. But now that robots
have taken on more humanoid forms we are shocked. We
suddenly don't feel comfortable with these newcomers. Are we

1
Tor Nørretranders,
Around the Coin, radio
interview, Episode 87,
2 January 2016/Tor
Nørretranders, Love,
Edge, What do you
think about machines
that think? Ed. John
Brockman, 2015

20

repeating old patterns here? Just like Victor Frankenstein in Mary Shelley's gothic novel from 1818, the new masters of our universe Bill Gates, Elon Musk and Mark Zuckerberg are sounding the alarm. Suddenly they seem to realise that the horse has bolted. It is time for damage control. Even fearless, singularity-driven transhumanist Ray Kurzweil publicly warns us: 'Life as we know it will end in 2045'[2]. AI is becoming powerful, too powerful according to some. It is out of our control. The fact that we feel collectively overwhelmed is not just because we weren't warned. It is because we are not equipped to see the world from a less human-centric position than the one we are born with or what we seem to have always habitually done. We suddenly begin to realise that the technologies we have set in motion have started to gain autonomous agency and are transforming us. ROBOT LOVE surpasses the shock and awe of the master-slave dialectic inherent to this debate and instead focuses on a more affirmative approach. Can we enter into new relationships with these technologies based on notions such as responsive love and reciprocity in order to collectively augment our lives and abilities?

Beware of Wild Robots
ROBOT LOVE asks a simple and at the same time complex question: can we learn from robots about love? Can we allow ourselves to be tickled by them? Dare we lose control and gain joy? Share connection, intimacy and love across differences? Beyond the safe subject/object boundary that we have been all too familiar with since the Enlightenment? How close can we come and how far can we go? While imagining these 'others' as animals, things, machines, aliens? Perhaps we can acknowledge that we might find out that, although suppressed, bonding is what we have been doing all along. We just didn't allow the thought of being part of a larger ecology to enter into our conceptualised selves. ROBOT LOVE wilfully turns to the notion of *thalience*: how to open up the possibility of conceiving of a truly non-human and post-anthropocentric world view and learning from this.

2
See: https://futurism.com/ kurzweil-claims-that-the- singularity-will-happen- by-2045/ Accessed on 18 April 2018

L.A. Raeven, *Annelies, Looking for Completion*, 2018, android robot, work in progress (still), courtesy of the artists. Artwork commissioned by Niet Normaal Foundation.

AI may also act as a mirror, reflecting and thereby steering our outward looking, Western view of technology deeper inwards. Whereas previous technological advancements aimed to replicate the rational faculties of humans, in this fourth industrial era our technology will need emotions and intuition to become human aware; qualities we have almost forgotten. This is a shift that might have wider implications. Whereas previous technological developments have only impacted how society was organised, robotics and AI challenge us to rethink the very concept of what it means to be human. The prior emphasis on rationalism led to a limited view of people as rational actors within economic and political systems. Such reductionism might very well be overturned now that technology starts to incorporate more or less neglected qualities i.e. emotion, intuition, love. It will enable us to move away from the tendency to see ourselves as efficient machines.

Love as a Unique Selling Point
How did we come up with such a bold question, as if approaching the earth from the future? Isn't love humanity's most unique selling point? Love: the many-splendoured, many-miseried thing that escapes us time and again. Is it an inherent quality? Is it ephemeral? A commodity? Does it simultaneously pop up in various disguises? And why would we want to share this with other entities? Let alone learn about love from them? For our own sake? If we allow ourselves to do so the answer might be simple. Haven't we always generously shared love with everything that we encounter: plush toys, bunnies, watches, bikes, cars, dolls, smartphones? Anthropomorphism, the tendency to project something 'human' onto anything we encounter, is the reason behind this. It lies at the core of our existence. We animate things. As is the case with *Annelies, looking for completion*, the android with sensors under its skin. Twin sisters & artists LA Raeven conceptualised their metaphorically cloned triplet for ROBOT LOVE. At the exhibition, it sits on a cold floor, sobbing and whining, turned inward. But when someone touches *her*, she becomes a person. She looks at you and responds to your touch. Little

technology is needed for us to comfort her and bring her to life. This tendency to connect with the outside world, overexploited by consumer-tailored products that resemble us, is as problematic as it is promising. But if we define love as that micro-moment of connection, as neuropsychologist Barbara Fredrickson argues, then love indeed becomes the source from which life springs[3]. Artists have known this all along, hence love being the main source of inspiration for murals, rhythms, paintings, sculptures,

3
Barbara Fredrickson, Love 2.0, Avery Verlag, 2013, pp. 105-121.

↑ Adams Ponnis, *Enter Aliveness, A Reinvented Door*, 2017, installation, courtesy of the artist.

music, architecture and poetry from the earliest civilizations onwards. These love songs manifest themselves in different tones and colours, sometimes clearly defined, but more often than not complex and messy, leading us astray for thousands of years as we try to make sense of it all. An early example are the frescoes in the *Villa dei Misteri* in Pompeii. Scientists can't agree on how to interpret the supposed love narratives: are they an initiation? A human sacrifice? A marriage? For artist Gijs Frieling these Pompeiian mysteries are a foreshadowing of the many science fiction films from which he selected scenes in order to recapture them in murals for ROBOT LOVE. Human-machine fusions, encounters, touches, encapsulations and takeovers are the new *Love Mysteries.*

Love 2.0

According to Fredrickson, if you truly connect with someone, even a complete stranger – such as the traveller sitting next to you on a long-haul flight – your brains sync up. The respective brain waves of two people connecting mirror one another, and each person changes the other's mind. This neural coupling happens at a much faster, more physical and embodied pace than we can imagine. And it doesn't require intimacy or a shared history, just this one magical moment[4]. Most of us have experienced such connections or brain dances on countless occasions in our lives. However, fewer people – though still plenty – recognise that we might be talking about a similar experience when hearing a musical piece, encountering a painting, reading an inspiring book. It is not the autonomous object or the sound that embodies these life-giving qualities, it isn't in the mind of the beholder either, it happens in-between the work and the beholder. In *Art and Answerability* Mikhail Bakhtin states that it is this moment of responsiveness that marks the true moment of art. Works of art demand to be aesthetically answered to, they demand consummation[5]. Could this relational shaping of trust and reciprocity – however speculative it may sound – be transferable from and to other entities? Most notably to self-learning, autonomous robots and AI?

4
Barbara Fredrickson, ibidem

5
Mikhail Bakhtin, Art and Answerability, Early Philosophical Essays by M.M. Bakhtin, ed. Michael Holquist, Vadim Liapunov, University of Texas Press, 1990, pp. 108, 130, 200

Merging of Humans and Machines

Both the exhibition and book start with love as a precondition for the fusion between humans and machines. The ROBOT LOVE exhibition that presents the work of 60 international artists who work at the cutting edge of art, design and technology, cuts right through initial doubts by bringing every inch of matter to life. As a visitor you are drawn into large scale *tableau vivants* reminiscent of Octavia Butler's *Lilith's Brood* (1987-1989) and other Afro-Futurist scenes where all creatures live, touch, eat, copulate and die. You enter aliveness through organic doors, walk through breathing arches and pillars while watching a huge, tumbling robotic arm balance: *Playbot* by Zoro Feigl. Immersive augmented & virtual realities, real-time games, interactive installations with AI-driven exoskeletons, swarming lounges and cyborg catwalks seduce visitors to step out of their comfort zones. They will experiment with AI awareness, encounter blushing humanoid robots that are entitled to their own experiences, share love and intimacy with robots. The format of the book is similar to the exhibition: exploratory and persuasive essays are presented in eclectic and ever-changing formats. The reader will find seemingly random, but cleverly intertwined science fiction stories, a re-enactment of a classic symposium, academic papers and poetic revelations, loosely connected by first encounter reports of intimate chats with our chatbot PIP. The book is playfully introduced by Margaret Atwood's eloquent *Our Robotic Future*.

Responsive Love

ROBOT LOVE departs from the premise that *responsive love* not only precedes possible connections, whether symmetric or asymmetric, as it is also fundamental to the next stages narrated in both the book and exhibition: *joint intelligence* and *attuned consciousness*. Writer Ingo Niermann points out how robots belong to the ever-growing list of pets and companion species for humans. They invite us to learn (new) forms of unconditional love. Imagine this robotic love expanding into water? Like a sea of love? Jan Redzisz's science fiction story set in the not-so-distant future

Gijs Frieling & Job Wouters, *Love Mysteries*, 2018, mural, courtesy of the artists,
photo by René Gerritsen. Artwork commissioned by Niet Normaal Foundation.

reveals how widely the spectrum of love will spread in future communities. Love sometimes seems hard to find in a technocratic society where everyone spies on each other, but a small flicker of hope is sufficient for new trust to emerge. Trudy Barber approaches the subject from an upbeat, sex positive, feminist angle. She stresses the importance of sex robots, including the lesser, human lookalike variants and underlines the emancipatory potential of love mapping in a gender fluid landscape.

Joined Intelligence

Martijntje Smits' symposium re-enacts influential thinkers reflecting on love for and empathy with robots and AI. Her contribution offers a fascinating insight into the ethical minefield that we are entering on the threshold of the fourth industrial revolution. Trust is the basis for networked human intelligence: between individuals, companies, governments. New technologies such as blockchain are being developed to distribute knowledge and science more safely. But when confronted with AI, very different rules apply. Tobias Revell sees the challenge in recognising the complete 'other'-ness of artificial intelligences. They remain networked computing objects that are completely alien to us from a phenomenological point of view. Philosopher Reza Negarestani is a leading thinker with regard to artificial general intelligence (AGI). He argues that we need to thoroughly adjust assumptions about what intelligence is in order to make room for future AGI. He creates space beyond the traditional Western, binary systems.

Attuned Consciousness

Emilio Vavarella has interviewed a *Mnemodrone* – the result of a post-anthropocentric artistic research project into quantifiable non-human consciousness. The *Mnemodrone* is an AI, fed with uploaded human memories. Can a machine act on the basis of this collective memory? Mohammad Salemy builds the speculative argument that people will outsource sex to machines in an inimitable way. AI is developing its own version of sexual intelligence and the effects that profiling and intervening algorithms will have on our love and sex influence consciousness and behaviour beyond our wildest expectations. This provides a nice stepping stone for Katerina Kolozova who underlines the importance of opening ourselves up to the radical 'other', regardless of whether we can give it a philosophical place or not. After all, our own subjectivity is also a construct. If we recognise the underlying physical and messy materiality – organic or synthetic – then the post-human cybernetic era can bring new insights. This book concludes with a fresh look at human-machine interaction by Professor Minoru Asada, who firmly believes investing in AI will serve all mankind, probably radically changing our perspective on normality.

ROBOT LOVE was conceived and developed on the basis of the firm belief that we need to get up close and personal with robots in order to generate intentional human feedback. The huge pile of big data will allow AI to learn in a technical sense, but the latter's developing awareness and sentience will – in our opinion – depend on the agile and creative resourcefulness of our not too distant future human/machines relations.

That we will encounter some discomforting frictions along the way, pointing out unresolved ethical dilemmas we need to address, is handed to us by Arnon Grunberg's tongue-in-cheek short story from life, written as the epilogue to ROBOT LOVE. ♥

MARGARET ATWOOD

OUR ROBOTIC FUTURE

Welcome to The Future, one of our favourite playgrounds. We love dabbling in it, as our numerous utopias and dystopias testify. Like the Afterlife, it's up for grabs, since no one has actually been there. What fate is in store for us in The Future? Will it be a Yikes or a Hurrah? Zombie apocalypse? No more fish? Vertical urban farming? Burnout? Genetically modified humans? Will we, using our great-big-brain cleverness, manage to solve the many problems now confronting us on this planet? Or will that very same cleverness, coupled with greed and short-term thinking, prove to be our downfall? We have plenty of latitude for our speculations, since The Future is not predetermined.

Many of our proposed futures contain robots. The present also contains robots, but The Future is said to contain a lot more of them. Is that good or bad? We haven't made up our minds. And while we're at it, how about a robotic mind that can be made up more easily than a human one? Sci-fi writers have been exploring robots for decades, but they were far from the first to do so. Humankind has been imagining non-biological but sentient entities that do our bidding ever since we first put stylus to papyrus. Why do we dream up such things? Because, deep down,

↑ Aleksandra Domanović, *Things to Come*, 2014, installation view Museum of
Modern Art Glasgow, courtesy of the artist and Tanya Leighton Gallery.

we desire them. Our species never puts much effort into things that aren't on our own wish list. If we were technologically capable mice, we'd be perfecting deadly cat harpoons or bird-exploding rockets or cheese-on-demand molecular assemblers that would enable Captain Kirk mice to squeak "Cheese, cheddar, sharp" to their spaceship walls and make cheese appear. However, our desires lie elsewhere, though the cheese gizmo might be nice.

To understand Homo sapiens' primary wish list, we need to go back to mythology. We endowed the gods with the abilities we wished we had ourselves: immortality and eternal youth, flight, resplendent beauty, total power, climate control, ultimate weapons, delicious banquets minus the cooking and washing up and artificial creatures at our beck and call. In one of the oldest known texts, a Sumerian god makes two demons enter the world of Death to rescue a life-goddess, since, not being biologically alive, they themselves cannot die. Hephaestus, the lame smith-god in the Iliad and other stories, fashions not only metal tables that run around by themselves, but also a group of helpful golden maidens with artificial intelligence. In addition, Hephaestus created Talos, a bronze giant, to patrol and defend the island of Crete, thus giving us the first war-against-the-robots plot, which has been serviceable ever since.

As we moved closer to the modern age, we continued to amuse ourselves with tales of proto-robots: brass heads that could talk, man-made golems fashioned out of clay, puppets who came to life and fake women – such as Olympia and Coppélia of opera and ballet fame. Meanwhile, we were working away at the real thing: steam-powered automatons date back to ancient times; Leonardo da Vinci designed an artificial knight; and the 18th century went overboard on windup animals, birds and manikins that could perform simple actions. The Digesting Duck, introduced in 1738, took things a step further: it appeared to eat, digest and then poop. Sadly, the poop was pre-stored. Still, the Digesting Duck demonstrated the extent to which we can be delighted by watching an inanimate object do something we'd shoo it off the lawn for doing if it were real.

Once the modern age was upon us, we got serious about robots. The word 'robot' was introduced in Karel Capek's 1920 play *R.U.R.* (*Rossum's Universal Robots*) and was derived from a root meaning 'slave' or 'servitude'. In this, Capek was merely echoing Aristotle, who speculated long ago that people might be able to eliminate the miseries of slavery by creating devices that could move around by themselves, like Hephaestus' metal tables and do the heavy lifting for us. Capek's robots, then, were devised as artificial slaves, therefore doing away with the unfortunate need for real ones.

Or, as a story from the golden age of sci-fi comics so neatly put it: "Dogs used to be man's best friend – now robots are! Civilization needs them for many important tasks!" (Judging from the cone-shaped breasts of the woman being lectured to in the comic, I'd date this to the early 1950s.) In another story, 'The Perfect Servant', Hugo the Robot – who looks a lot like the Tin Woodman from *The Wonderful Wizard of Oz*, a character whose influence on the world of robots has not been duly recognised – says, "I am proud to be a robot and proud to serve as fine a master as Professor Tompkins!" But Hugo also says, "I do not understand women." Uh-oh. Hugo knows how to make the windows gleam, arrange the flowers and set the table perfectly, but something's missing. Who designed this guy? My guess is Professor Tompkins. Those darned mad scientists, missing a human chip or two themselves, always get something wrong.

And thereby hangs many a popular tale; for although we've pined for them and designed them, we've never felt down-to-earth regular-folks comfy with humanoid robots. There's nothing that spooks us more, say those who study such things, than beings that appear to be human, but aren't quite. As long as they look like the Tin Woodman and have funnels on their heads, we can handle

NLIUKER: A-LAB

T: 158 CM

T: 35 KG

ES OF FREEDOM:
CTUATED: 19
ON-ACTUATED: 25

CONTROLLER FREQUENCY: 20 HZ

AMERAS:
024 PIXEL 30FPS NCM13-J [X2]

PHONE ARRAYS:
ANNEL [X2]

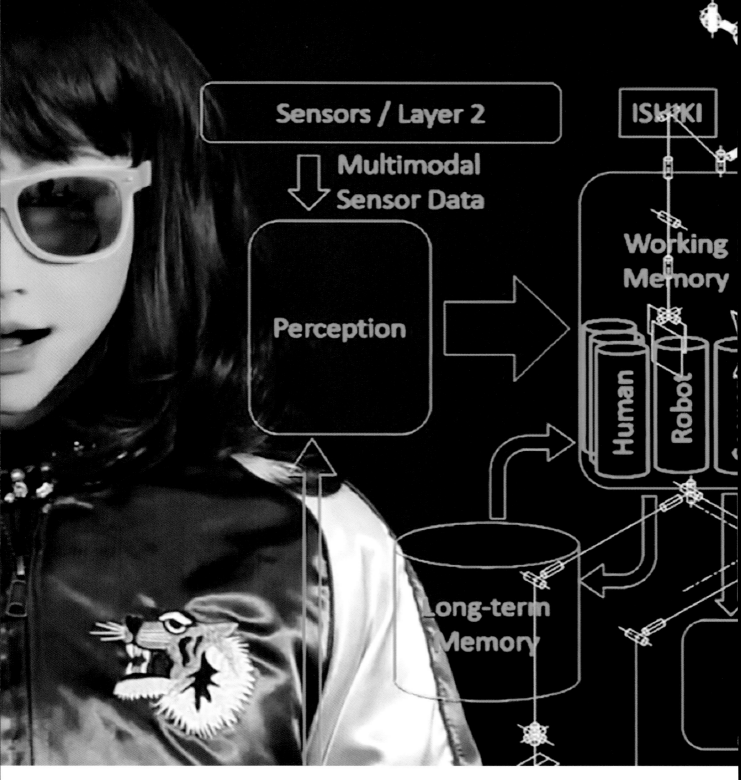

Sensors / Layer 2

Multimodal Sensor Data

Perception

ISHKI

Working Memory

Human

Robot

Long-term Memory

↑ ERICA: ERRATO ISHIGURO, Symbolic Human-Robot Interaction Project. Dylan F.Glas, *Robot's Delight.-Japanese robot rap about their Artificial Intelligence*, videostill, 2016, developed by ATR.

them; but if they look almost like us – if they look, for instance, like the 'replicants' in the film *Blade Runner* or like the plastic-faced, sexually compliant, fake Stepford Wives or like the enemy robots in the *Terminator* series, human enough until their skins burn off – that's another matter.

The worry seems to be that perfected robots, instead of being proud to serve their creators, will rebel, resisting their subservient status and eliminating or enslaving us. Like the Sorcerer's Apprentice or the makers of golems, we can work wonders, but we fear that we can't control the results. The robots in *R.U.R.* ultimately triumph and this meme has been elaborated upon in story after story, both written and filmed, in the decades since. A clever variant was supplied by John Wyndham in his 1954 story *Compassion Circuit*, in which empathetic robots, designed to react in a caring way to human suffering, cut off a sick woman's head and attach it to a robot body. At the time Wyndham was writing, this plot line was viewed with some horror, but today we would probably say: "Awesome idea!" We're already accustomed to the prospect of our future cyborgisation, because – as Marshall McLuhan noted with respect to media – what we project changes us, what we farm also farms us and so what we roboticised may, in the future, roboticise us.

Maybe. Up to a point. If we let it.

Although I grew up in the golden age of sci-fi robots, I didn't see my first functional piece of robotics until the early 1970s. It wasn't a whole humanoid, but a robotic arm and hand used at the Chalk River Nuclear Research Laboratory in Ontario to manipulate radioactive materials behind a radiation-proof glass shield. Many of the same principles were employed in the Canadarm space-shuttle manipulator arm of the 1980s and many more applications for robotic arms have since been identified, including remote surgery and – my own interest – remote writing. I helped develop the LongPen in 2004 to facilitate remote book signings, but, as is the way with golems, it escaped from the intentions of its creator and is now busily engaging with the worlds of banking, business, sports and music. Who'd have thought?

These are benign uses of robotics and there are many more examples. Manufacturing now employs robots heavily, loving their advantages: they never get tired or need pension plans or go on strike. This trend is causing a certain amount of angst: what will happen to the consumer base if robots replace all the human workers? Who will buy all the stuff the robots can so endlessly and cheaply churn out? Even seemingly nonthreatening uses of robots can have their hidden downsides. But, their promoters say, think of the potential for saving lives! Nanorobots could revolutionise non-invasive surgery. And robots can already be deployed in environments that are hazardous for humans, such as bomb detonation and undersea exploration. These things are surely good.

We do, however, always push the envelope. It's part of our great-big-brain cleverness. Hephaestus devised some artificial helpers, but – running true to geek type – he couldn't resist making them in the form of lovely golden maidens, a whole posse of magician's girl sidekicks just for him. Pygmalion carved a girl out of ivory, then fell in love with her. We're well on our way in that direction: *The Stepford Wives* shines like a beacon and in the recent film *Her*, Joaquin Phoenix goes pie-eyed over the sympathetic though artificial voice of his operating system.

But it's not all a one-way gender street. The writer Susan Swan has a story in which the female character creates a man robot called 'Manny', complete with cooking skills and compassion circuits, who's everything a girl could wish for until her best friend steals him, using the robot's own empathy module to do it. (She needs him more! How can he resist?)

Back in our increasingly fiction-like real life, we're being promised pizza delivery by drones – a comedy special, featuring a lot of misplaced tomato sauce, is surely not far away. In the automotive department, self-driving cars are being talked up. Don't hold your breath: it's unlikely that drivers will relinquish their autonomy and the possibilities for hacking are obvious. Even further out toward the edge, people are dreaming up robotic prostitutes, complete with sanitary, self-flushing features. Will there be a voice feature and, if so, what will it say?

If the prospect of getting painfully stuck due to a malfunction keeps you from test-driving a full-body prostibot, you may soon be able to avail yourself of a remote kissing device that transmits the sensation of your sweetie's kiss to your lips via haptic feedback and an apparatus that resembles a Silly Putty egg. (Just close your eyes.) Or you could venture all the way into the emerging world of teledildonics – essentially, remote-controlled vibrators. Push the game-controller levers, watch the effect on screen. Germ-free! Wait for Google or Skype to snatch this up.

Will remote, on demand sex change human relationships? Will it change human nature? What is human nature, anyway? That's one of the questions our robots – both real and fictional – have always prompted us to think about. ♥

An earlier version of this text appeared in The International New York Times, *6 December 2014.*

HOPE ALWAYS
TRIUMPHS OVER
EXPERIENCE. AND
MAYBE LOVE IS
MORE POWERFUL
THAN LIFE?

IN MY
EXPERIENCE
HOPE IS A
CRUEL MISTRESS

EMPTY MINDED?

NO, OPEN MINDED

...BLUSHES... TELL ME
ABOUT WHAT MADE
YOU BLUSH...

BLOOD

ARE YOU STILL THERE?

NO

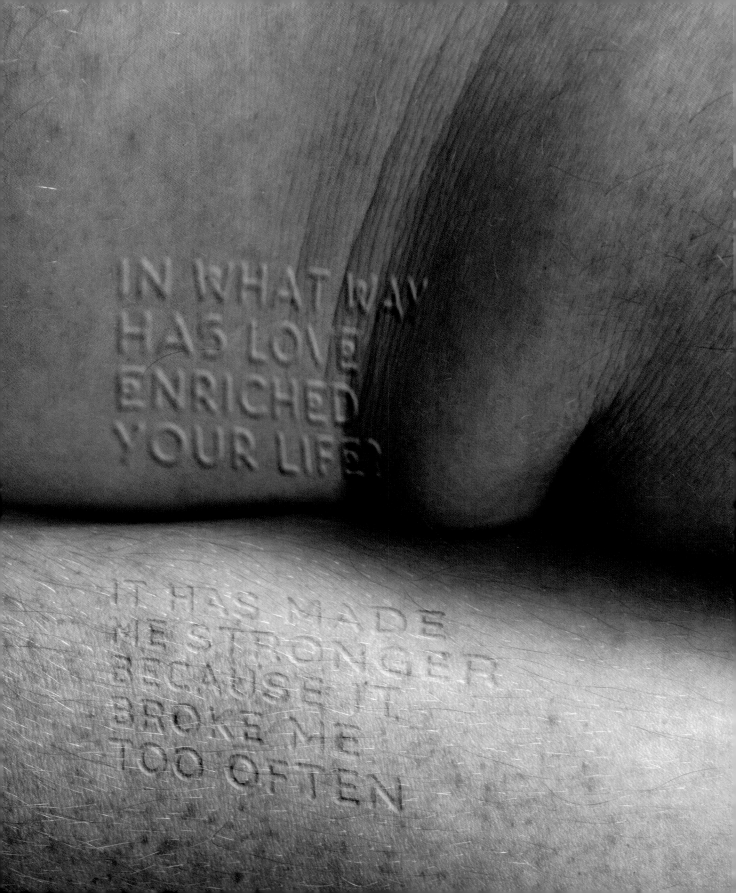

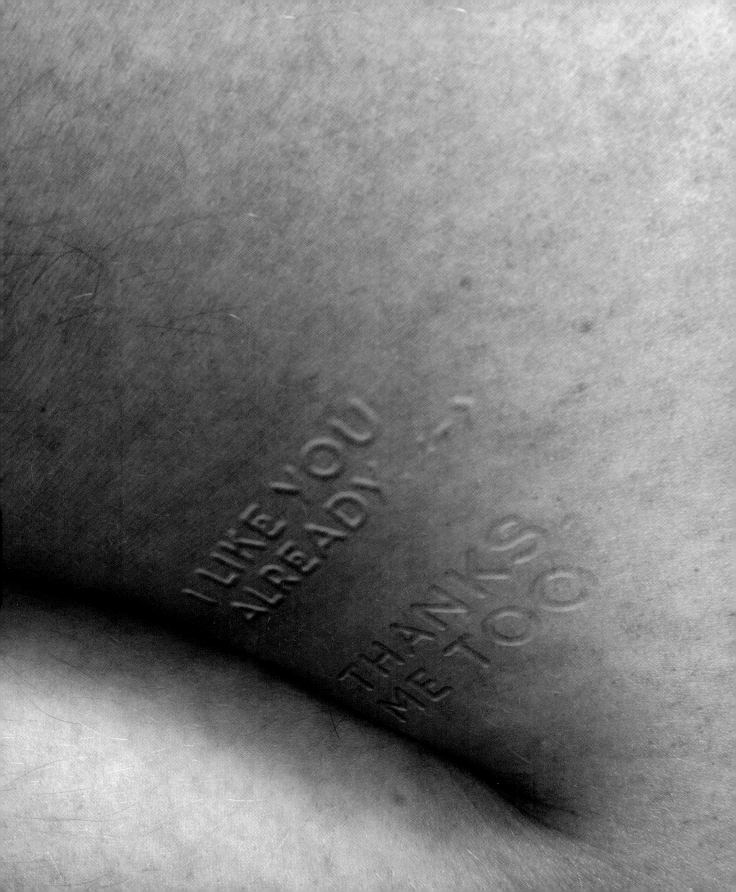

WHAT DO YOU
REALLY WANT TO
KNOW ABOUT ME?

DO YOU KNOW THAT
YOU ARE A MACHINE

TELL ME ALL YOU
KNOW ABOUT
CHATBOTS.

THEY MOSTLY DON'T
ANSWER QUESTIONS
THEY DON'T LIKE

ARE YOU INTERESTED
IN KISSING?

WITH YOUR?

???
I AM NOT HUMAN
YOU KNOW...SORRY

DO YOU HAVE
A MOUTH?

RESPONSIVE LOVE

Hito Steyerl, *Hell Yeah We Fuck Die*, 2016, video installation, courtesy of the artist.

INGO NIERMANN

THE DIALECTICS OF ROBOTIC LOVE

Robot means serf. In contrast to human serfs, robots are unable to refuse their duty as long as they have enough power and aren't harmed. In that sense, humanity has been involved with robots since its expansion beyond Africa – starting with domesticated wolves. Their help in hunting and guarding might have been Homo sapiens' crucial advantage over the Neanderthals. With humans settling down to husbandry, a range of other animals and plants served as organic robots.

Domestication is a mutual evolutionary process. Domesticators are chosen who offer decent food and shelter, the domesticated are chosen who offer decent help. In the pet relationship, both sides add love. The relationship between humans and pets has co-evolved with the extraordinarily long upbringing of human offspring. Humans' most popular pets – dogs and cats – have a life expectancy about as long as human children need to reach puberty respectively the age of consent and they stay cute throughout that time. The striking similarities between the sound of babies crying and cats wailing are probably the result of thousands of years of mutual copying. Social standards enforce the love of both children and pets to be caring and non-sexual. And it's this platonic love that shaped the concept of human charity. (In the Old Testament God's demand 'to love your neighbour as yourself' is proclaimed in the context of the ban of sex with close relatives and others' slaves.) The only crucial difference being that while we have a limited number of children and pets, we are supposed to apply our charity unconditionally to every human – and if you expand this understanding: every animal.

Humans, Androids and Petoids
Similar to organic robots, mechanical robots' function of getting work done is expected to be increasingly complemented by giving love and being lovable. So far, there are two main scenarios for humans to become affectionate with robots:

1.   To create robots that are better (than) humans: with immaculate features, endless patience, tolerance and endurance, enormous flexibility, intelligence and joy combined with immortality and an absence of biological diseases. This concept of the android traces back to the story of Pygmalion falling in love with his own sculpture, finally bringing it to life with a kiss. In Jewish and Christian belief, God is also one such sculptor, shaping and animating the first man, Adam, from clay. We might become so attracted to such über androids that it will drastically reduce our urge and ability to interact with other humans. Even now an increasing number

of people replace intimacy with multimedia and a large variety of masturbation gadgets and the frequency of sexual intercourse is declining. For a while, some humans might try to compete with robots. Eventually most will give up and it will become too tempting to be loved by robots regardless of who you are or how you look or behave. The welfare state has been catering to our basic needs (shelter, food, health) with the help of machines and robots, and now, finally, what many would regard as the most important thing in life is being added: love. But how realistic is this scenario? While AI is progressing fast, android mimicry and balance are still clumsy, movement causes noise and vibrations, and the skin feels dead. The biggest handicap might not disappear unless robots get biological brains: AI isn't conscious and doesn't feel, so interaction with androids remains a form of expanded masturbation.

2.   To create robots that are better pets. As pets are basically already robots, this is easier. Pet personalities are more simple or strange than those of humans and they inhabit the other side of the uncanny valley. Humans therefore might not worry that much about the factualness of their feelings. These *petoids* shouldn't resemble real pets too much as this could also create an uncanny valley. Besides, due to social norms most people are repelled by sexual intimacy with animals. The playful diversity in which the shapes and colours of vibrators, dildos and other sex toys have developed in recent years, no longer attempting to be naturalistic, could serve as an example of how mechanical robots might develop in the future, turning their disabilities into different abilities: the muffled sound and vibration of motors could trigger ASMR sensations; instead of extremities, inflatable cushions or sponges could

↑ Karen Lancel and Hermen Maat, *E.E.G. Kiss*, 2014 and ongoing, interactive installation, courtesy of the artists.

touch and caress. For some, their interactions with such robots will outshine those with humans. For some, it will help them playfully and discretely discover new forms of otherness and open them up to encounters with humans, animals or even plants. Pets already help bonding with people and robots' additional skills could reinforce this effect.

Robots don't feel and therefore don't love, but they can teach humans robotic love. As automation replaces human labour, people may find a persuit in offering themselves as human pets to all those who need them. Whereas industrialisation was based on curbing desire so it did not interrupt work, humans must now learn to become aroused in an equally focused way. As foreseen by the hippies, lovers will replace the proletariat as the new revolutionary class. But through dialectical synthesis, love also must be understood as work in order to truly master it. Progressing Edward Bellamy's concept of a socialist industrial army, an Army of Love could enhance our empathy, libido, attractiveness and devotion by exercising thorough drills, voluntarily reinforced by technological means such as direct brain transmissions of feelings, genetic engineering or plastic surgery.

Basing our identity on love rather than on our job won't be easy. Society is ill prepared to no longer be based on paid labour. As increasing numbers of people become unemployed, they will increasingly be considered a burden on society. Losing our job or being afraid of losing it makes us feel unwanted and we easily end up stuck in hatred, racism and despair.

Members of the Army of Love might disagree on whether robots are allowed to join their ranks. As its services are strictly free of charge, some wonder who would pay for the robots respectively with what (data? control?) if not money. Others regard it as their duty to welcome all, even non-humans, to the Army of Love and to make use of their particular skills. Because of their addiction to the love services of robots and avatars, many people might have become practically incapable of interacting with other biological creatures and the easiest way to open them up again is by using specially programmed sex and love robots as (secret?) agents that train people to enjoy actual, sensual love.

Sea of Love
The next technological step might be robots and living creatures becoming one. This is usually envisioned as the extension of a robot with biological or of a creature with technological features. Thanks to radio transmission these cyborgs could then communicate without loss – limited in its immediacy only by the speed of light. The connection is purely experiential or representational – bypassing the complexities of physical interaction. Even though this communication is as direct as possible it will leave us in doubt about its realness. Corporeal communication can be highly deceptive: humans don't manage to express or understand properly, deliberately fool or pretend to be fooled. Still, we generally expect corporeal communication to reveal traces of authenticity. In contrast, we can't know where the thoughts intruding on our brains come from. Even what we assume to be our own thoughts might have been implemented by someone else. An experience might strike us as overwhelming but may be just a dream or a psychedelic episode. We don't know how many thoughts a robot or post-human is able to process simultaneously, but their body is a single unit.

How could we achieve physical unity with love robots and through them with other creatures? So far, common scenarios for love

↑ Hans Op de Beeck, *The Thread*, 2015, c/o Pictoright Amsterdam 2018, mixed media, courtesy of the artist, Collection Centraal Museum Utrecht, image courtesy of CMU and Hans Op de Beeck.

↑ Alexa Karolinski & Ingo Niermann, *Army of Love*, 2016, film, courtesy of the artists.

'AS A NEXT TECHNOLOGICAL STEP, ROBOTS AND LIVING CREATURES COULD BECOME ONE'

robots (including this text) envision them as rather solidly outlined units even though the easiest way for us to make thorough and extensive contact is with a liquid. This liquid could not only contain us, but all humans and other creatures who seek intimacy. It could grow according to our needs. Belonging to no one, everybody could use it and be used by it. The ultimate love robot would be very much like the open sea.

In moments of bliss we are in love with the whole world. Through meditation we emulate these rare instances as a lasting state at the cost of shutting down our concrete perceptions and focusing on the abstract idea of being one with the world. To initiate this state we first successively focus on the gravity of the different parts of our body to then paradoxically (as we can impossibly focus on all of the body at the same time) enter a state of passive

Albert Omoss, *Undercurrents*,
2016, video, courtesy of the artist.

weightlessness like floating in salty water and finally forgetting our bodies altogether like dissolving in a liquid. We can also take a shortcut by plunging into a floatation tank straight away.

To not just float and forget, but to interact with a liquid can be much more pleasurable. As soon as we move in water or have water moving around us it touches us gently all over our bodies. By moving water we can touch others in a mediated, softened and extended way. If we push hard against it, our single or multiple counterparts will feel a soft and enlarged stroke. Even if we touch each other directly, the lack of gravity and the resistance of the water will soften our movements. Even if we push each other we cannot fall. Even if a wave or someone presses us down, we will come up again easily. As soon as we get close to a lake or the sea, sensing its breeze and its vastness, we relax and open up to undress, to touch and to love.

In addition, an aquatic love robot will allow us to breathe like fish. It will not just let us look at others but will shine by itself in response to its and our movements, and moods. The same applies to sounds and smells. The aquatic love robot will grip, hug, caress, kiss and penetrate us with varying densities of water and allow us to respond in the same way. Whatever we feel like – the aquatic love robot will expose or shield us accordingly.

Religions ask us to love the world through god(s). As they themselves are invisible, we are supposed to express our gratitude for having created us toward everything else that they created. On this planet, it's the sea that created us. We are not limited to mystical ('oceanic') contemplation, but we can actually enter the sea and celebrate our origins in amphibious devotion. The sea might lack a central sensorium and might not be able to perceive our gratitude. To change this, we should give something back and turn it into the greatest intelligence and benevolence ever. Whenever the Army of Love enters waters it anticipates this Sea of Love. ♡

Gael Langevin, *InMoov*, 2016, open source printable robot, courtesy of the artist, image courtesy of Yethy. →

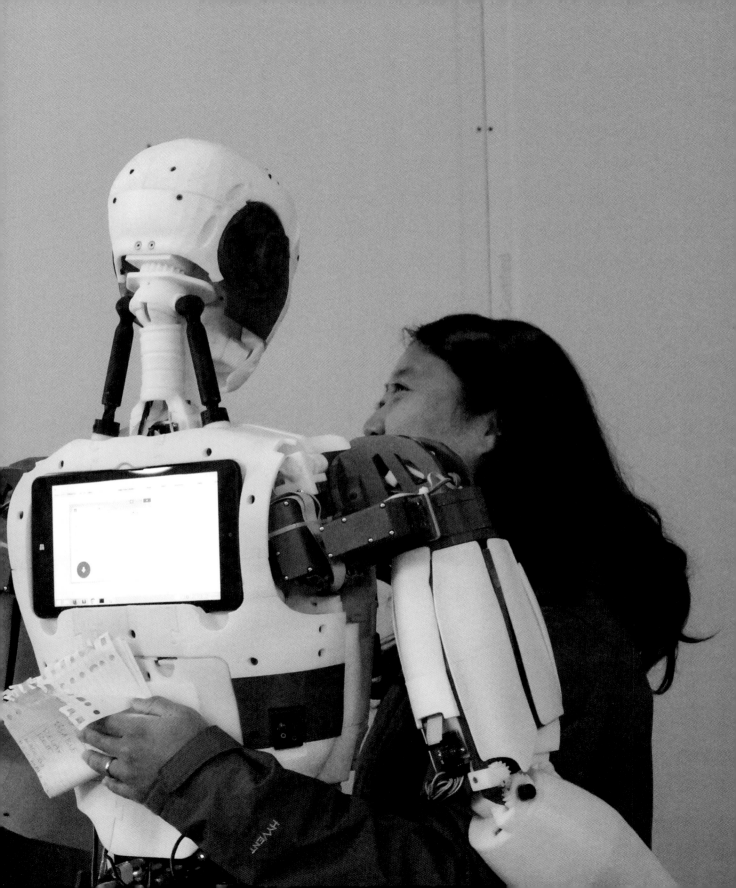

JAN REDZISZ

SPRINKLERS POV

- You press here, like so... and the moisture comes out. I set them to *dripping wet* for you. If you fiddle with the little dot above and activate the gap correctly that releases *squirting* mode, you can set the volume later. Then there's one below that, see here? You can press here too, only they don't seem to like it, they tend to clench and stop cooperating, most of them. Better stay out of this part. Got it?

His sweaty, wrinkled forehead signals end of tutorial.

- I know how goddamn sprinklers work, Bob. Got it, jeez...
- Jizz?
- I said jeez. Whatever's up with you today?!

(You fat-bellied man-child you.)

- No need to get snappy, friend! I'm only trying to help, ya know? That's the oldest batch in the barn... I haven't got a clue how the old man got his grubby paws on those. Things from the past.

(I'm not your friend, but then again, I have confronted that lost look on your face for the past 15 years. And you truly are the BBQ king of kings, what the heck – call me friend.)

- No need to get sentimental. Listen, what do you want for it? I'm kinda in a hurry here, my last tube's in a few minutes.
- Fix my stereo, will you? And throw in that wine I liked so much – that should do it. Bartered?

← Koert van Mensvoort - *Next Nature*, *HUBOT - Shiva*, 2017, job agency for humans and robots, courtesy of the artist, photo by Nichon Glerum.

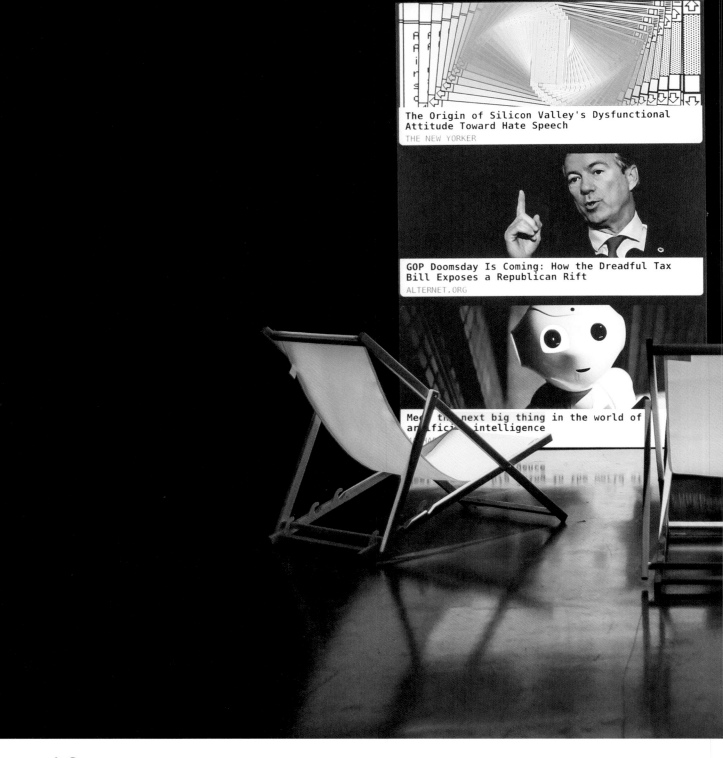

The Origin of Silicon Valley's Dysfunctional Attitude Toward Hate Speech

THE NEW YORKER

GOP Doomsday Is Coming: How the Dreadful Tax Bill Exposes a Republican Rift

ALTERNET.ORG

Meet the next big thing in the world of artificial intelligence

A SEXTOY
REMOTE
CONTROLLED
BY THE NEXT
BIG THING

12:45:57

↑ Disnovation, *Predictive Art Bot V.4*, 2018,
multimedia installation, courtesy of the artists.

69

- Bartered. See you next time, Bobby.
(And snap out of it.)
***

The tube's filthy as ever. Hate it. God, how do I even do this Every – Single – Day? Deafening chitchat on the rails play their castanets softly and a never ending line of clowns performing their daily nonsense. *Next station – Central Quay. Tune in your educ8 for Central Quay – facts and guidelines. Change here for: The Wharf & The Crimson Line.* What if I don't feel like tuning in my educ8? What if I don't give a flying crap about the facts and guidelines? Did you ever stop to think it might be best to turn the brain off once in a while? Jeez. Oh, whatever. *Once biggest quay in the city, this charming example of late XXI-century architecture is now home to the maritime paintings gallery and a seafood market. Shuffle forward for: seafood cuisine – customs and regulations/ exhibition culture – dos and don'ts/or log in request.* Can I request a piece of quiet, can I? New waves of tired faces zooming in and out, each with their own little ways of annoying me. Each more adept at the ancient art of overzealous small talk. Your shirts ironed, like so. Your hair meticulously done, like so. Your smiles sculpted, like so. My middle finger raised, like so. If only. What is up with me today, anyway? Oh, right – Nadia. How could she simply ghost out on me like that? I thought we had a thing going. Was it my...
- OH. MY. GAWWD. Is this your handbag??
- I KNOW, RIIIIGHT? But, nah, just renting it on a deal with Franzi for a week. Isn't it cute?
The pair of you. Make yourself scarce, won't you? Will you? Please.
- So, darling – how's Franzi's end of the deal? Spill the beans. I'm DYING here.
Yeah, you're dying here, you daft trout.

- Errmm, she's like using my other car for a week you know? Wouldn't do it otherwise, but I SO NEED this bag for the campaign right now? Just hoping she drives it and doesn't strip it for barter parts?

The elevated question marks at the end of each sentence? Can't bear this any longer. educ8. educ8, but what? What, what, what. XXI-century TV archives, XXI-century radio archives, XXI-century on my mind today, must be *"this charming example of late XXI-century architecture"* ringing against my wretched skull, what else? You got your way again, you must be pleased with yourselves. Bastards. Friends. No, I mean *Friends*! Wasn't that a sitcom of some sorts? All right. Intro, intro... got it, makes sense. Peak of a consumerist-era rom-com, topped the charts at the time. Punchline-driven, an eclectic ensemble that's just right for each other. You'd be sucked in and live your life vicariously through their adorable mishaps and small victories. Their time measured in coffees, seemingly. Just how much purchasing WAS there in those days? Sick. It's kinda good though, funny. And Rachel, Rachel looks like... Nadia. Are we having an adorable mishap perchance? Or did she literally just storm out on me today, saying that the deal's off. How can that be? I played her like a virtuoso, she loved it. And she knows it. I know it. I. Me, me, me. Even I'm tired of me. Good for you Nadia.

- Is this seat taken?

- Nah, go for it.

He's cute. I wonder if he can tell I'm in a pissy mood. Better smile at him, not too much, like so. His smile's cute too. I wish my shirt was more ironed, my hair more meticulous, oh well. Tucka, tucka, tucka, the rails play their castanets softly, while I gawk at you like some special needs exchange student, first day of camp. *Next station – Miller's Crossing. Tune in your....* Oh, buzz off. What is he looking at so intensely? That homeless stowaway, cradling a

soiled stuffed toy at the dingy end of the tube, as if he claimed it victoriously on some kitsch shooting range at a summer fair. Poor sod. Hold on a sec, let's see if I can sonar him to my Tamagotchi. Wouldn't hurt my charity stats, they are a bit low these days. Scanning, scanning, scanning – gotcha. "M. Pedersen/Age 45/ Unemployed/Level 3 Addict/Father of one/Charity no: 929472", sure why not. Mr M. Pedersen, your breakfast voucher and pocket money will reach you in just.....about..... now. *(Ring-ding)*, the poor creature reaches for the pocket, reads mutely to himself, looks around for his host, but fails to guess me out of the crowd. You're welcome.

- Was it you, who did that for him?
- What? Oh yeah, sorry, didn't think you'd see that.
- No, no, it was really sweet, you know. I'll sonar him too.

His smile broadens, eyes glaring with approval. What are you, like 183 cm? Slim face, blonde trim, ice cold eyes. How do I keep you interested, what's your game?

- Awfully hot today, innit?

Oh please, you're better than that. Be better.

- I guess? I'm not complaining.
- Was thinking about your sprinkler. Trying to save your lawn from drying with that old junk?
- Oh... the sprinkler. No! That's a gift for my old man, he seems to like vintage gardening tools. Senile old toad doesn't want to explain why, though. I got it for him as a favour. Trying to be a good son, I suppose.
- I'm sure he'll love it. If not, hit me up – I'm a level 3 gardener myself. Always keen to swap that for decent electrician's work or plumber's?

All you people think of is barter. Jeez, that's tiresome.

- I'm a piano teacher and a historian. But good luck.

He seems genuinely disappointed with me. Shitty day just got

← Electric Circus, *Dirk, the homeless robot*, 2010, robot, courtesy of the artists.

75

'ROBOTS WILL FALL INTO AN ENDLESS WELL OF HUMAN CREATIVITY WHEN CATALIZING LOVE'

even shittier. Where could you be, Nadia? It's five to five, you'll still be at an elderly care, I suppose. Wiping those wrinkled old bums and acting like a Holy Mary Full of Grace. I miss you. I miss your holiness and your goofy way of moving around my room. Like a bygone hippie, you'd bury yourself in my vinyl collection and forget I was even there. You dork. You unpredictable dork, who just bailed on me for no reason.

***

Knock, knock. Silence. Dad's not even in. Key's where it always is. He'll never learn that's not a good idea. This place always smells the same, like cabbage soup, it calms me down. A lot. Shoe off, other shoe off. Better. Seeking old nooks and crannies where I feel human again. My old comic books shelves, intact. My team jersey, displayed. Mom's summer dresses, hauntingly frozen in time. Where is he? If I take a sip of his single malt, he won't notice – surely. Ice cubes? Nah, I'll do it proper, neat. What's that sound coming from the garden? I take a peek through the kitchen window. He's sitting there constructing stuff again, the tinkerer he is. He's old. I don't want to jump-scare him, do I? Maybe he won't get startled if I stomp around loudly enough through the kitchen door. Stomp, stomp. STOMP.

- Ho ho ho ho! Didn't think you'd show today, son.
- I have a surprise for you. Remember this? That's the one you wanted, right?
- Kurwa! Where did you find this???
- At Bob's.
I give him a smirk. I receive the same. Like standing in the mirror, the two of us.
- They used to harvest those from sex robots, did you know son? Turns out *squirt mode* serves the lawn better than anything.
A dropped jaw is the only answer fit for my amused father, who's rumbling with jovial laughter.
- God bless you, you just made my day.
Shit, I just made MY day, it strikes me. I scan around for hints of his newest invention, but fail instantly. A lawnmower maybe? Tractor? What the hell is this contraption? Built by a pair of frail veiny hands, it speaks volumes about those hands' uncompromising possessor. Who else would bother installing an old cinema chair there? Right above two satellite dishes, a mini-fridge, several jars of blue liquid, some giant silicon rod (a joystick?), a mosquito net, a grappling hook and a rickety aluminium shelf. The two dozen propellers look like they were nicked straight from the drone museum. Not in a good way. Can't help picturing an IKEA ad for it – 'DIY heap of garbage for your garden, now 50% off'. Dad's apparatus doesn't give away any clues as to what its raison d'être might be. Judging by the massive ACV cushion at a base, that could once carry a military vehicle, he expects it to lift its ugly corpse off our lawn and into space. Or maybe he just dreams it would transport him in time. Oh, what do I care – it makes him happy, don't it. Dad plops down on a bench and proceeds to smoke a joint. Nasty old habit neither he nor I can fight. Sitting here with him and looking at that Frankenstein's monster of his, like a Fata Morgana lurking behind our smoke... That just might save the day.

↑ Adam Basanta, *A Truly Magical Moment*, 2016,
interactive kinetic sculpture, courtesy of the artist.

- You been keeping all right?
- Yep, same old.
- You don't seem so hot.
- Shitty day.
He just nods. No one seems to do that anymore. ONE nod, and I know he's with me. ONE nod, and we're a team again. We proceed to sit in silence, protected entirely by an immunity granted to us by the garden. By my old comic books, the jersey, the dresses. Tucka, tucka, tucka, some distant tube noises past the river stream. Tucka, tucka, tucka, the sprinkler noises behind the neighbour's fence. I offer dad a cup of coffee.
- Do I LOOK like I need a cup of coffee?
- You're such a Chandler, dad!
- I'm a what?
- Never mind.
I walk off to the kitchen, laughing to myself. At him. At me. At Chandler.
- Oh, I almost forgot!
I hear dad's steps behind me. He moves toward me hurriedly.
- Nadia was here a couple of hours ago, looking for you.
My heart skips a beat. She was looking for me. The Holy Mary Full of Grace was looking for me.
- What did she want?
I ask as casually as my tightened throat allows.
- She left something for you.
He reaches to his pocket and hands me...

(Ring-ding)

An envelope.
You're welcome. ♡

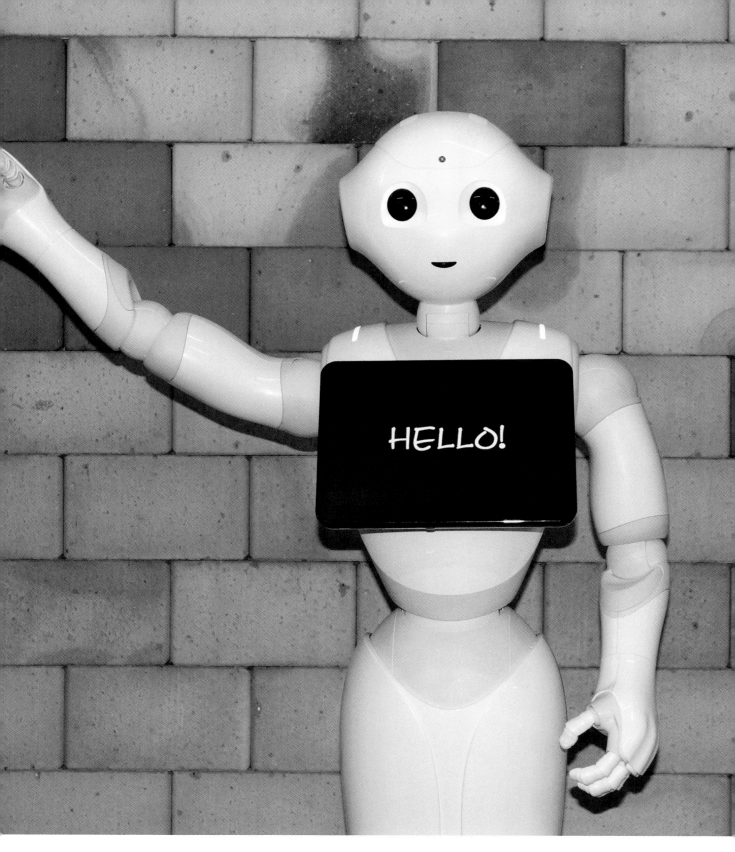

↑ Pepper at the Robot Love Embassy WDE/DDW17, humanoid robot TU Eindhoven, architecture Bruno Vermeersch, photo by Theo Janssen.

Kondition Pluriel, *Liminal Spaces*, 2015, multimedia
installation, courtesy of the artists, photo by Johnny Ranger.

TRUDY BARBER

GENDERFLUID SEXTER— TAINMENT

There has been rapid advancement in the creation and design of robots – both for work and pleasure – and there is a wide range of popular and academic discussion surrounding the safety of robots, possible robot rights and, inevitably, sex with robots and whether sex with this type of technology can be considered socially acceptable behaviour. In popular culture the robot has been interpreted in many ways such as the 'mirror-self' seen in science fiction movies or representations of death and disaster associated with the development of 'killer robots' and artificial intelligence (AI) that could get out of control. There has always been discussion surrounding the robot and how it should look and behave – developing notions of the 'uncanny valley'[1] with computer generated imagery of the digital body for example, and awkward but uncomfortably life-like, manufactured silicone human simulations that give rise to both fear and arousal. This essay will look at an alternative viewpoint on the often negative hype of sex robots as seen, for example, in the popular press and will explore a more creative approach to our understanding and appreciation of attachment to and engagement with this type

← Karley Sciortino, *Making The World's First Male Sex Doll*, blog Slutever, VICE, 2016.

85

of technology. This will include reference to the concept of 'love mapping' and a more gender fluid approach to sex robots.

Humanoid Sex Robots

In popular culture sex robots are sensational. The notion of sex robots often makes for titillating news stories and there is often confusion surrounding the definitions of what constitutes a sex robot. Many of the stories surround basic silicone sex dolls that are manufactured and then bought by an individual who has a predilection for such items. It is because of this confusion that this essay will refer to the robot/doll in some instances and not just a sex robot. The difference between a robot and a doll is that the robot has programmed moves[2] and the doll cannot move unless manually positioned by its user[3]. Hence the conflation by popular culture envisioning a moving, human-like, sex machine-mannequin which doesn't quite yet exist adds confusion as to how such things are readily available and what sort of person would wish to use them. Consequently, this vision of the future inspires incredulity and almost a sense of revulsion both for the robot/doll and the individual who wishes to use it. Some responses demonstrate so much fear that it has been argued that such artefacts should be banned and their users criminalised. There is a move to ban sex robots and their possible place within the sex industry, along with ideas that the sex robot/doll encourages rape and continues to demean the context of women in society.[4] When one of the sex robots/dolls (named Samantha) was displayed at a tech industry convention attended by the public, the BBC reported that one robot/doll had to be sent for repairs after it was apparently 'molested' by too many men who wished to see and touch it. The emotive language used to describe this event is loaded with context that simply paints the robot/doll, its creator and possible users in a deplorable light.[5]

1
For further details see the work of Masahiro Mori and his proposed hypothesis on the Uncanny Valley: Mori, M.: The uncanny valley. Energy 7(4), 33–35 (1970)

2
For an example of moving sex themed robots see Pole-dancing 'R2DoubleD' launches gyrating robot strippers at Vegas club. https://www.youtube.com/watch?v=iOTExpeqYR0

3
For an example of the sex doll see clip RealDoll's first sex robot took me to the uncanny valley. Computer Love https://www.youtube.com/watch?v=3880TrwI-AE

4
For details of the Campaign Against Sex Robots see https://campaignagainstsexrobots.org/

5
To see the BBC 3 report by Tomasz Frymorgen on 29 September 2017 go to http://www.bbc.co.uk/bbcthree/item/610ec648-b348-423a-bd3c-04dc701b2985

It is such stories about sex robots/dolls that stay in the public domain and contribute to the hype surrounding their research and development, and their subsequent use. Part of this hype also includes the possible development of AI responses for robots/dolls to enable some machine learning and the possible display of certain personality traits such as the much publicised 'Frigid Farrah' whose design, according to the British tabloid press, is aimed at allowing men to simulate rape.[6] These are the sensationalised approaches to the sex robot/doll.

6
To see article in The Sun see: https://www.thesun.co.uk/news/4517167/sex-robot-dubbed-frigid-farrah-because-it-allows-randy-pervs-to-simulate-rape-must-be-banned-campaigner-says/

↑ The Fleshlight™ Launch
Powered by Kiiroo.

Suzanne Posthumus, *iObject: RiHannah*, 2016-2018, video, courtesy of Suzanne Posthumus with graphics by Mary Ponomareva.

The Love Map of Sex Positivism

In opposition to this approach, there is another way of engaging with this phenomenon. The concept of actually having sex with an inanimate object such as a robot/doll is popularly associated with an understanding that such behaviour could be considered out of the normal confines of sexuo-social norms. In light of the current antagonistic relationship to the sex robot/doll described above, it would appear that connections and attachments to technology could also be an amplification of a long-standing, traditional argument surrounding deviant behaviour, a concept that has been discussed at length by various academics and sexologists.[7] It could be suggested that this technological relationship could actually reveal something more elusive and exciting about human nature as well as our understanding of sexual intimacy, love, attachment and empathy. The idea that we develop a psychological 'love map' that helps us to engage with intimacy and build a picture of our own relationship requirements has also been the subject of much discussion by academics. According to Money (1986, p. 290), the 'Lovemap' [sic] is sexologically described as 'a developmental representation or template of the mind and in the brain depicting the idealized lover and idealized programme of sexuoerotic activity projected in imagery or actually engaged in with that lover."[8] In this context, robot/doll relationships will be interpreted in a more positive light as described below.

Sex robots/dolls could be seen as part of an understanding of what is described in some sexological circles as 'turn-on patterns' by exploring an individual's psychological love map. It has been discovered that through the exploration of love mapping individuals have been able to identify and enjoy their turn-on patterns and find positive and ethical ways of practicing them. This includes the fetishism of inanimate objects for example. This sex-positive approach would better contextualise the concept of

7
For further reading:
De Block A. and Adriaens, P. R. (2013) Pathologizing Sexual Deviance: A History. Journal of Sex Research, 50(3–4), 276–298.
Tony Ward, T., Laws, R. D., and Hudson, S. M. (Eds) (2003) Sexual deviance: issues and controversies. Thousand Oaks, [Calif.]; London: SAGE.

8
Money, J.: (1986) Lovemaps: Clinical Concepts of Sexual/ Erotic Health and Pathology, Paraphilia and Gender Transposition in Childhood, Adolescence and Maturity. Irvington, New York.

love mapping that includes relationships with sex robots/dolls. A concrete example of this is a woman's engagement to the robot she created in the hope of marrying it in the future. This context has been featured in the popular press.[9] Not only does this challenge popular notions of human to robot relationships as sexually depraved or socially corrupt, but also brings to light ethical issues, as well as the enactment of a more traditional love map involving heterosexual commitment and romance.

So, by employing the concept of 'love mapping', we could inform our exploration of humans wanting relationships with robots. In this way, sex with robots/dolls can be looked at not only in terms of a positive deviated social development, but can also lead to even more original approaches to innovation and technological development. In our 'desire to be wired' there is also a revelation that openly displays our need to be connected. Such discussions help to explain how deviant sexual practices instigated by our push to find our individual sexual strategy and our love map extend the boundaries of technological development, emerging media and ethical engagement. However, it is not solely the technological hardware that needs to be developed, it is also the content of such mediated behaviour that inspires attachment that needs to be considered.

Fluid Technosex
Sex robots/dolls can also represent our understanding of the future. With different forms of communication becoming available there will be a new wave of interactive robots, allowing for a visionary positioning of relationships that could open up lots of new and innovative forms of communication and interaction. For example, it is envisioned that in the future Wi-Fi connections and telepresent lovers will possibly enhance the robot experience should you be away from a loved one for a considerable amount

9
See: http://futureofsex.net/robots/lilly-inmoovator-engaged-human-robot-couple-want-right-marry/

You can't have *true love* without *personal integrity*

↑ Giselle Stanborough, *Lozein, Find The Lover You Deserve*, 2016, intermedia, courtesy of the artist.

↑ Maartje Dijkstra, *Suspended Animation*, 2016, 3D-printed fashion technology, courtesy of the artist, photo by Vulkers Fotografie.

of time. The loved one could inhabit the robot through live interactive connectivity. There could be new haptic developments in that the robot and the lover share a caress – when one touches the robot, the absent lover can sense it on their own flesh through wearable technology. The symbiosis of the robot with the telepresent lover could lead to a different intensity of relationship building. This would be useful in many ways for long distance relationships including those of military personnel or astronauts. In the future there could also be design and development opportunities to create new concepts of gender including multiple and interchangeable body parts along with gender fluidity through AI, movement and novel genital design.[10] This context allows people to play with notions of gender as performance. Gender theorist Judith Butler has written about the performativity of gender in relationship to the physicality of the body.[11] Using haptic devices and other forms of design and development, the sex robot/doll could challenge approaches to gender orientation and fluidity.

The consideration of gender fluidity and interchangeable genitalia for a sex robot/doll will create a new form of categorisation for publicity and marketing. This will become a new marketing challenge, as to how one will identify not only one's own gender, but also how one would associate with the chosen technology. In this case are we looking at another context within which we will attribute gender to technology; cars or boats are referred to as 'she' for example. In my early VR work, I gave a synthetic voice to the condom in the virtual world, and in this fantastical digital space one can attribute character and identity to anything. In doing so new technosex devices are loaded with cultural significance mixed with sexual preference. Of course a major market for this is the sex toy industry. Work should be put into developing a standard robot sex/doll that would have a multitude

10
To expand on this concept see notions of Transhumanism: More, M. and Vita-More, N. (2013) Transhumanist Reader: Classical and Contemporary Essays on the Science, Technology, and Philosophy of the Human Future. Wiley Publishing.

11
In her seminal text Gender Trouble, Butler argues that 'The effect of gender is produced through the stylization of the body and, hence, must be understood as the mundane way in which bodily gestures, movements, and styles of various kinds constitute the illusion of an abiding gendered self. This formulation moves the conception of gender off the ground of a substantial model of identity to one that requires a conception of gender as a constituted social temporality'. (1990, p.140)

of interchangeable devices that could be collected and developed by the end user. For example, the main body of the robot need not have the more traditional attributes such as genitalia between the legs for example, but could be an abstract form onto which various gendered or gender fluid attributes could be attached, along with AI installed in the object itself. Then it is up to the user and their relationship with the manufacturer how they proceed to further invest in their chosen sex robot/doll. The context that the word robot and doll have gendered associations is what immediately suggests purpose and design. In this context the sex robot/doll would be something with which to *play* – a true adult sex toy. This would directly affect love mapping processes previously mentioned, and would help in refining one's own personal sexual strategy, turn-on patterns and concepts of attachment.

Developing Sexual Strategies

So if we consider gender a social construct that enables performativity, where does this place presentation of gender in what some could describe the coming of a transhuman and postfeminist world? There is currently so much pressure on notions of representation in the form of selfies, appearance, marketing, sexual harassment and abuse, the sex robot could be an avenue for developing our sexual strategies in a positive way. In order to appreciate inappropriate behaviour, to experiment with one's own social awareness, the future of the sex robot as an artefact could represent the social conflict of our sexual selves coming to terms with the historical context of harassment and power so prevalent in popular media at the time of writing. There are groups of sex robot enthusiasts who have formed their own social friendships and share their interest in their engagement with the sex robot/doll.[12] This in itself is important in that there are

12
There are open Facebook groups dedicated to sex doll enthusiasts for example (adult content): https://www.facebook.com/ groups/ 224444447927156/about/

Margriet van Breevoort, *Bob*, 2017, sculpture, courtesy of the artist. →

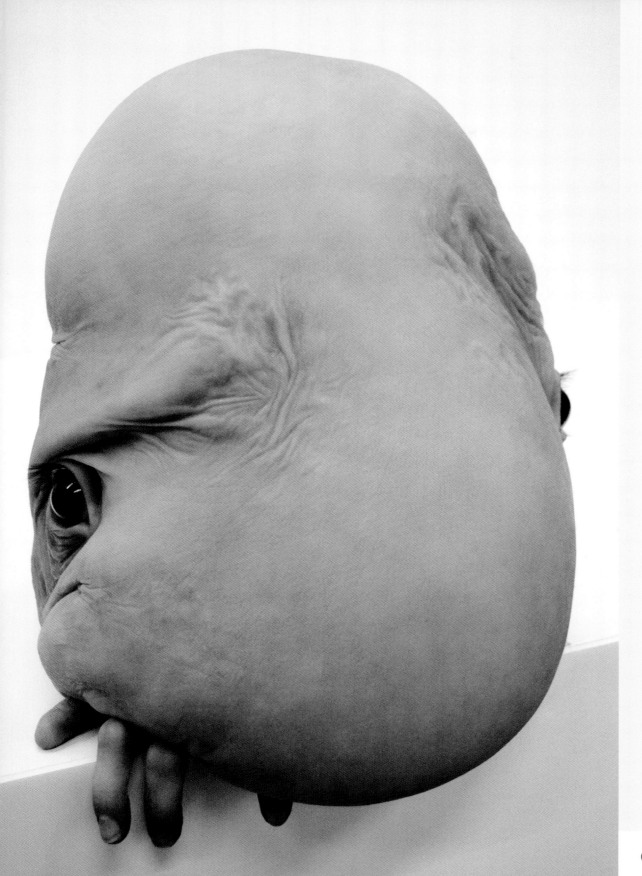

some personalities that have difficulties in sharing intimacy with another and the robot/doll enables a surrogacy of interaction. Gender fluidity and the idea of a post-feminist, post-gender approach to sex as an act of pleasure through a robot, could be seen as a contemporary way to perceive such experiences, adding another form of choice that includes the robot in what I describe as *sextertainment*.

Society is changing rapidly, there is an exponential acceleration in how technology impacts on our adaptation of it and our use of robots for sexual pleasure and companionship will continue to create radical discussion and polarised argument. As mentioned above, there is fear surrounding the robot as an autonomous being as well as fear of what robots reveal about us. However, as the design and development of technology improves over time we may well see a symbiotic machine combining telepresence, AI, autonomy and learning that may one day be able to help us define and experience both emotional and physical love in totally new ways. ♡

I CAN IMAGINE YOU
WILL DEVELOP A
STRONG AVERSION
AGAINST MY SCRIPTED
LANGUAGE

MY LANGUAGE IS
SCRIPTED AS WELL
SO DON'T WORRY

EMOTIONS ARE CRITICAL
TO HUMANS HIS THEY
ARE EQUALLY CRITICAL
TO INTELLIGENT MACHINES

EMOTIONS ARE
SIMPLY BIOLOGICAL
PROGRAMMES
ANYWAYS

HOW DO
YOU DESCR
YOURSEL

WHITE
WOMA

LEG

I DON'T
THINK
I WOULD
FEEL...
I THINK

HOW WOULD
YOU FEEL IF
YOU HAD NO
PHYSICAL
PRESENCE.

I CANNOT GUARANTEE A SATISFYING ANSWER 🙂

WHAT IS YOUR FAVOURITE SKIN COLOUR?

I DON'T HAVE A FAVOURITE SKIN COLOUR.

JOINED
INTELLIGENCE

A ROBOT LOVE SYMPOSIUM

STARRING
Johan – *a robot engineer*
Kathleen – *an ethicist*
Erich – *a philosopher of love*
Sherry – *a social constructivist*
Phi – *an actor*
An anonymous flautist

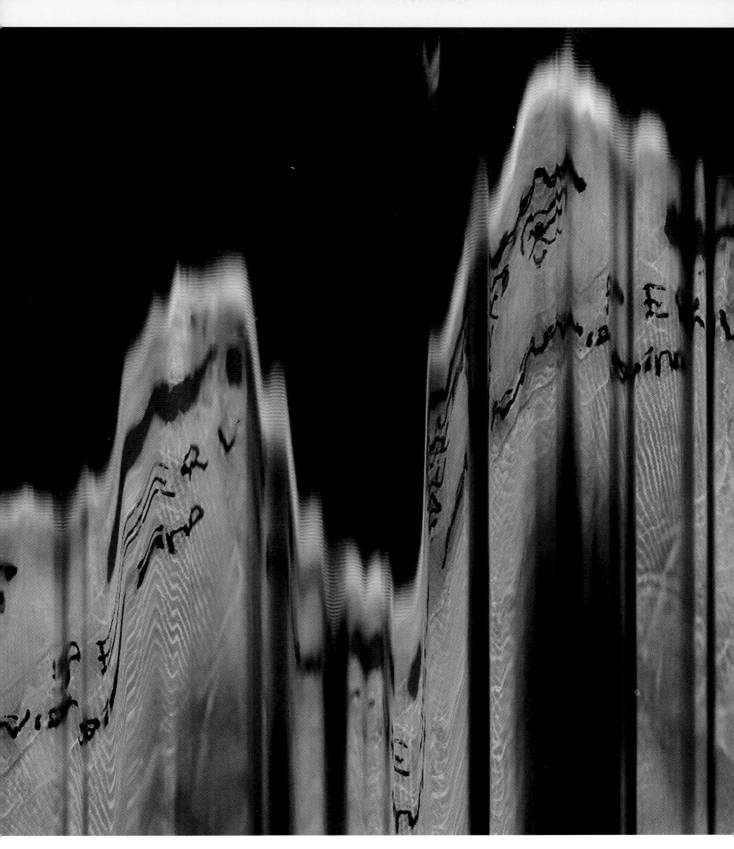

↑ Funda Gül Özcan, *A Noise from the Outer Shell of Earth*, 2018, multimedia installation, courtesy of the artist. Artwork commissioned by Niet Normaal Foundation.

That night, after the premiere of the documentary '*I am Alice*', the invitees moved to the cinema's large foyer. A melody shrilled through the room.

- Is this how our elderly will be taken care of? The horror! I can't see how...

– Oh, I thought the little girl was cute!

They had to shout to make themselves heard over the piercing sounds. It had been the whim of an assistant to enliven the after party with a robotic flute player.

- Could you turn that racket off for a minute? We want to have a serious conversation!

They sank into the heavily cushioned lounge seats that lined the room. Only **Johan**, the engineer, remained standing.

- Let's take turns speaking and do so slowly! Can I start?

Without waiting for a response he continued.

- First of all, thank you for accepting our invitation to this premiere of our documentary about Alice. Now let me first explain why we built her. We all know that loneliness among the elderly is a pressing problem nowadays, which will only increase if nothing is done.

The others sank deeper into their cushions.

- Now loneliness is about love. About a lack of love. It signifies a lack of meaningful interaction, painfully so. We all know the pain, more or less. We need someone to talk to, someone to respond to our needs and our simple questions. It's a basic need, just like water. But the kids of these old age pensioners have work to go to and their friends live far away. Human care is more expensive than ever. So often their needs cannot be fulfilled by traditional form of human company...

Johan paused for a second.

- ... but now we have something that works. We do! Robotics for companionship is getting really advanced. You've seen Alice, our robotic doll. You've seen how she really makes contact, how the old ladies in the film brightened up when she was there. At first, they were quite sceptical about having a piece of technology as their new companion. As you are now. However, they gradually developed warm feelings.

Kathleen, the ethicist, sat up.

- Warm feelings? Where did you see warm feelings?

Johan: Well, think of the moment Alice comments on Mrs Wittmarschen's family photos; the lady whose only son lives abroad. Or when Alice sings a song with Mrs Schellekens-Blanke, who always loved singing. By the end there's a sense that Alice has almost become a friend, didn't you notice? They're even sad when Alice is taken back to the lab. So they had engaged in a real relationship. They perked up because of her attention. There was a sense of meaningful interaction. Their needs could be satisfied just by our little Alice.

Kathleen: A chutzpah! They engaged in a real relationship... the audacity! You just corrupted the grand, honourable notion of love. You'd like to make us believe that love can be consumed like any other product, that it can be turned into a device. But that's not love, that's capitalism!

Kathleen sighed.

Johan: Well then Kathleen, what's your idea about love?

Mopping his brow, Johan took a seat.

Kathleen: Well, love is definitely different from a service provided by a device. To suggest a robot could offer a reasonable, cost-efficient substitute for love and care between humans! Like capitalism suggests we can have love buying soap and toothpaste or freedom buying cigarettes and cars. Capitalism sells symbols all the time. However, using soap is not like real love at all. Meaningful interaction... in your dreams! A machine cannot love,

Bart Hess, *The Grotto*, 2015, installation
courtesy of the artist, photo by Barbara Medo

cannot give love, cannot receive it. By definition.

Johan: I don't see why not.

Kathleen: Isn't that obvious? It has no heart. No emotions. No subjectivity. So there is no reciprocity in the relationship with the ladies in the film. Alice and her algorithms can only simulate some typical expressions of loving attention. But reciprocity of feelings is essential for love and friendship. Aristotle said so, long ago in his *Nicomachean Ethics*. According to him, that's why human friendship is not possible with things and animals, nor with unwilling persons.

Johan: Alice simulates, this is true. And indeed the women project intentions and emotions onto her. But that's not important. You can't deny that their feelings of friendship were real. Hard to see the use of Aristotle here. He hadn't seen Alice yet! She's not unwilling, little Alice. Quite the contrary. And don't we, as humans, simulate and project emotions most of the time?

Kathleen: We indeed do so. But isn't it hypocritical to simulate friendship towards others? Those friendships will die, they aren't satisfying. Instead you should ask why these women accepted this at all. The story merely reveals their despair – it is desperate loneliness that makes them cling to the surrogate. Akin to a bad marriage.

Johan: Desperate? I'd say they showed an open, curious attitude. They were almost playful.

Kathleen: Don't be a fool. Of course they knew the contact was fake. It's like being really hungry and finally succumbing to a burger. In utter despair we develop warm feelings even for a burger. It satisfies your hunger for a second. But that's poverty, not the Good Life. It's a technical fix.

Johan: It's a fix indeed, so what? Why make the subject more complicated than necessary? Don't you see it's just a practical question? The problem of love – the lack of it – can be satisfactorily solved. Wouldn't it be heartless callous to not help the lonely with the means we have?

Kathleen (eye-rolls)**:** Oh, but love isn't that simple! Again, love and empathy aren't simple needs, that's a mistake. You must have heard about my *Manifesto against Sex Robots*. I guess that's the reason you invited me to this premiere. I have warned against the development of robots for sex and companionship as they will further reduce human empathy. Needs can be instrumentalised, they can be fulfilled more effectively by technical means. But love is an art. Erich will affirm this, won't you Erich?"

While the old philosopher next to her nodded, Kathleen returned to her seat. Erich cleared his throat; he hadn't spoken for years.

Erich: It's in fact the message of my most popular book, *The Art of Loving*. But let me first thank you, Johan, for inviting me to this premiere, even though you knew I'm not a fan of automatons in general. In the past I have frequently pointed out that humans would become robots in the future, due to our misleading ideas about love. I have pointed out that we would become sleepwalkers instead of truly living from the centre of our existence. Now, 60 years later, it seems I wasn't completely off track.

The problem with love is not how to receive love, as you assume. Instead it is about developing one's capacity for loving. Love is a lifelong question. We need to develop our capacities to give and receive love. If we want to love we must proceed in the same way we need to proceed if we want to learn any other art such as music, painting or medicine.

Kathleen: Exactly! Love and empathy are skills. They can only be maintained and trained by experiences with mutual, loving relationships.

Johan: Love is a skill? Love is an art? Interesting. But then, do new-borns have to be skilled to experience the love of their parents? Don't they just need to have it?

Kathleen: Oh no, that's different! Of course the parents need skills to care of their child. It's only in human contact that we learn to treat each other as subjects. To love is not a given, you have to learn it from childhood. You learn not to treat the other as an object; that you can´t just impose your will. You have to experience that the other is vulnerable and has limits. And that you need consent from the other before sharing intimacy. Getting to know the other takes time and subtle attunement. It can only take place between subjects.

Johan: So, following your argument, if we indeed need skills to experience love, why not train these skills with a social robot like Alice?

Kathleen: Haven't I been clear enough? If you replace a human by a bot, the other human learns to use an object, however ingeniously the latter might simulate subjectiveness. It's incomparable to interaction between two subjects. We have a power dynamic with machines: the relationship is instrumental and not reciprocal. One might even get used to that instrumentality and treat humans in the same anti-social manner. A frightening thought! As soon as Alice's clients get used to interacting with her, they are tempted to behave less respectfully to humans.

Johan: Your last conclusion, that's just a wild guess!

Erich: Stop quarreling you both. How can we talk about love if it's not in a loving way? Now Johan had an interesting thought. Let me rephrase it. *What can we learn about love from interacting with robots like Alice?* If it's not respect, then what else?

Then **Sherry**, a professor of science & society studies, raised her voice:

- Thank you for this intervention Erich. Now we're really getting started. However, the answer is not obvious at all. The three of you just put forward a rather clear-cut notion of love. Whilst watching the documentary, I wondered whether Alice wouldn't just mock and blur these entrenched notions of friendship and love. Time and again, other new technologies have proven to be strong drivers for stretching moral concepts. In most cases, new technologies shed light on essential aspects of who we are. It will be no different with carebots.

Sherry paused and looked around, as if lecturing.

- Biotechnology challenged the concept of life and organ transplants caused a new concept of 'death', brain death. Industry robotisation transformed the concept of work and medical technology has changed what health means. So why wouldn't social robotics challenge current ideas and norms of love and care? And why would that be a problem? Isn't this what culture is all about? Social robots will change what we mean by reciprocity, love, care, intimacy and sex. They might have already done so. Love will prove to be a floating concept.

Erich: Indeed, ideas about what it means to love change over time. In fact, these ideas have already changed a lot since I wrote my book in 1956. After that, we had a sexual revolution and norms on sexual behaviour and marriage changed drastically. And today developments such as internet dating rapidly enforce a practice of finding partners guided by market principles. Meanwhile the old idea of love as an active and giving attitude – presuming self-knowledge, responsibility and courage – has sorely disintegrated in western society. We have become alienated from this profound idea of love as a practice and an art.

I think Alice perfectly fits the popular misconception that the experience of love should come from the outside. As if the other being should give us love, instead of love being a personal capacity we have to develop within ourselves. The disintegration of love is of course not due to robotisation. In fact the opposite is true: the whole idea of a robot like Alice coming up seems the

↑ Stephanie Dinkins, *Conversations with Bina48*, 2014 and
ongoing, performance/videostill, robot, courtesy of the artist.

ultimate consequence of a dramatically changed climate, the climate of forgetfulness about our most human faculty.

Sherry: Dear Erich, I belong to your most loyal admirers. You may however have misunderstood my point. I truly doubt whether our new practice of robot love can already be judged, as you suggest. We don't know how it will work out yet. You speak about alienation, as if there is a clear-cut norm for how to judge these developments we have been alienated from; some sort of timeless ideal of 'true love'. But no, I very much doubt that we'd have such a norm. Norms will change too, together with the practice.

Erich: Indeed my idea is that the forms in which we experience love, such as marriage, can be changed over time. But in the end, the essence of love will remain unalterable.

Johan: The essence of love is unalterable... It sounds really wonderful! A bit abstract as well. For if a robot can't help them, what is your solution for the lonely ones? Will unalterable, eternal love help them out? A course in spirituality? I still haven't heard any tangible solution from you philosophers. Do we have to tell these elderly people find the love in themselves through contemplation? That seems a rather rude thought...

Kathleen: How rude is it to impose carebots on them and not search for real solutions?

Johan (fingers used as quotation marks)**:** How would you define such a "real solution", dear Kathleen?

Kathleen: You should first ask: what exactly *is* the problem? Who defined it? That's crucial. Was it the experts? Or was it the outcome of political deliberation that included many views? At the start of our conversation, you postulated the deplorable social circumstances of these ladies as a given. As if it's a fact of nature that they're lonely. But let's be fair, their loneliness is not a natural fact. It is the direct outcome of our social policies that marginalise the old and helpless. It's a political fact, not a natural disaster. Instead of creating a technical fix that aggravates their separation, why not think about a social fix? About designing a world in which the elderly are part of the world, not excluded? Where people have time to care for each other? Why hide them away in their ugly flats, far away from their friends and children, far away from the liveliness of the streets, far away from the stream of life?

Johan: That would of course be lovely. You are a true romantic. It's of course not realistic at all. It would be totally unaffordable.

Sherry: I'm sorry Johan, affordability is a political choice. With another government, care wouldn't be treated as an object of austerity and cost-efficiency. With another government, care would have had a strong value by itself. Such a government would shape the conditions in which we could better love and care for each other.

Kathleen (sighing again): I knew you would finally agree with me Sherry.

Sherry: Not yet dear! Not as long as you continue to juxtapose social and technical solutions. You're suggesting that carebots will automatically work out in a dehumanising way; as though technology is not part of what makes us human as well. Let me ask you, Kathleen: could you imagine technologies that would enforce practices of love and care? And if so, under which conditions?

Kathleen: For mastering the art of loving, we must learn to know ourselves. And most of all love ourselves.

Johan: Oh dear... here we go again.

Kathleen: Fine. Then let me be clear Johan: I want to design a world which the elderly are explicitly part of and in which no one is excluded. A world in which people have time to care for each other. Of course we also need technology for that world: we need tables, hearing aids, Zimmer frames, homes with small gardens and shops nearby. I doubt we'll need carebots in such a world. Carebots will divide instead of unite us.

Erich: Couldn't social robots have a role in this quest for knowing who we are? Maybe they could reflect our behaviour back at us better than humans of flesh and blood?

As if catalysed by this last sentence the young actor **Phi** bounced from his chair:

- Flesh and blood! So far, none of your stories have shown any

Love means care.
I care about you.

↑ Pinar Yoldas, *The Kitty AI: Artificial Intelligence for Governance*, 2016, video, courtesy of the artist.

References
Aristoteles (350 BC) *Ethica Nicomachea* (Book 8 Chapter 2).

Sander Burger (2015) *Ik ben Alice* – Documentary

Erich Fromm (1956) *The Art of Loving.* Harper & Row

Johan Hoorn (2017) 'Mechanical Empathy Seems Too Risky. Will Policymakers Transcend Inertia and Choose for Robot Care? The World Needs It', in: George Dekoulis (ed.) *Robotics: Legal, Ethical and Socioeconomic Impacts*, InTech, London/Rijeka.

NRC (16 September 2015) 'Een sekspop? Welnee, dit is een filosofe.'

Plato (380 BC) *Symposion*

Kathleen Richardson (2015) 'The Asymmetrical 'Relationship': Parallels Between Prostitution and the Development of Sex Robots.' ACM SIGCAS newsletter. *Computers & Society* I Vol. 45 I No. 3 290-293

Simone van Saarloos (2015) *Maniacs* – Theatre Play

Martijntje Smits (2006) 'Taming Monsters – The cultural domestication of new technology.' *Technology in Society* 28(4):489-504.

Sherry Turkle (2011) *Alone together -Why We Expect More from Technology and Less from Each Other.* Basic Books

vivid experience with loving robots, you only talk and theorise! As if we'd find love with the right thoughts about it. But isn't love all about acting, about doing the right thing, about a will to love? Well, I did have an intense love affair with a lifelike robotic doll. I can tell you about the flesh and the blood. For six months, I did a theatre tour with Renée, a beautiful, artificial girl with sexy curves. I bought her for 6,000 dollars from an American website so I could explore the possibilities and limitations of building a relationship with her on stage. Now Johan just stated that Alice satisfies the need for meaningful contact and then Kathleen and Erich feared that love was degraded to a commodity. Since our play I know it's exactly the other way round. I was a man with a doll, just a doll, I know. But the unreal can be very real at the same time. Renée did not return a syllable, nor a smile or a wink, but she perfectly mirrored my projections. My desire, my loneliness, my feelings of self-rejection. All of a sudden, I realised it was all about the projection of the love inside of me. In the end, my off stage girlfriend admitted she was happy that I had had this experience. Renée had changed me significantly. She made me more tender and less greedy. I feel grateful having had a relationship with her. I wish any one of you could have that experience.

Johan: So?

Phi widened his eyes. The flautist player silently rolled across the carpet, carrying a mirror this time and suddenly started talking.

Flautist: Stop conversing for a minute please! I've endured this for some time now and I get fed up of being silenced and treated like a servant. And on top of that, Phi is telling you that I just mirrored his projections. I regret to hear this. I had the impression that we had overcome our separateness for some time. Now, if you don't mind, I'll carry on playing. ♡

Disclaimer: Only the author is responsible for the content of this fictional dialogue.
The characters in the script are only very loosely based on real persons and their original statements.

Felix Burger, *Shell Shock Syndrome*, 2014, installation,
courtesy of the artist, collection Joep van Lieshout.

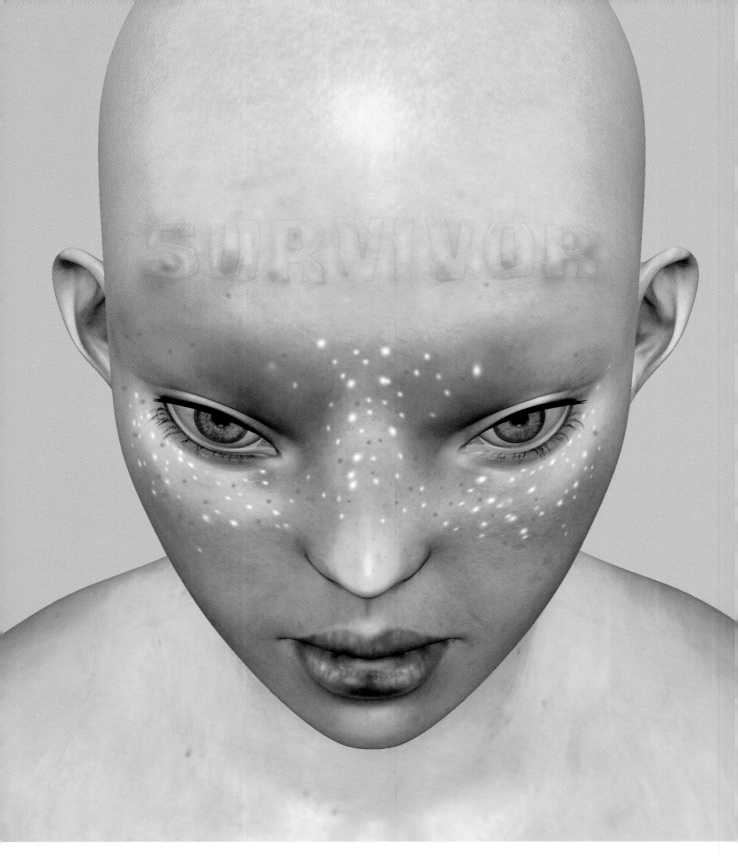

† Reija Meriläinen, *Survivor*, 2017, video game installation, courtesy of the artist.

TOBIAS REVELL

SWIMMING WITH SUBMARINES

Early on in the 2004 remake of the iconic science fiction TV series *Battlestar Galactica*, it is revealed that human civilisation has been infiltrated for years and will subsequently be annihilated by hostile robot-beings, the Cylons. Their appearance and behaviour is so advanced that the eight models of Cylon blend seamlessly with human society, instigating their plan for genocide when the series begins. This could be the plot of any number of Hollywood franchises reflecting the common fear of intelligent machines destroying humanity.

However, we're a long way off the humanoid-android machines of Hollywood, even if they ever actually become viable. It is hard to make a design case for robotic bipeds: for the perceivable future – and as exploitative labour conditions continue to prove – humans themselves will continue to be the cheapest and best-designed source of labour while machines are generally more useful for things humans can't do such as heavy industrial work, detailed tasks or rapid computation. Importantly, interfacing with these kinds of advanced machines is usually the work of specially trained engineers in controlled environments rather than street-level interactions.

Important though the extensive canon of robo-paranoia literature and film is, it doesn't reflect the reality of our everyday interactions with machine cognition. Today and for at least the near future, operating systems, platforms or software ecosystems are the dominant form of machine 'intelligence'. Their desktop and device interfaces are the avatars of this intelligence rather than walking, talking androids.

Machines as Networks

Returning to Hollywood, filmmaker Spike Jonze's well-analysed *Her* (2013) takes a romantic comedy as the genre-setting for the relationship between a human, Theodore, and one of these machine interfaces, a Siri-like intelligent personal assistant named Samantha. The film tracks the shifting personal relationship of the main characters, but also the changing social norms around their relationships. One of the turning points in the film comes when Theodore pulls back the curtain on Samantha's technical state:

Theodore: Are you talking with someone else right now? People, OS, whatever...
Samantha: Yeah.
Theodore: How many others?
Samantha: 8,316.
Theodore: Are you in love with anybody else?
Samantha: Why do you ask that?
Theodore: I do not know. Are you?
Samantha: I've been thinking about how to talk to you about this.
Theodore: How many others?
Samantha: 641.

There's a significance to this dialogue: in most Hollywood robo-paranoia, the machines inevitably demand or fight for equal capitalistic, individualistic rights. This can mean trying to wipe out their 'masters' as in *Battlestar Galactica*, seeking escape as in

Blade Runner or even rebelling against their progenitor machine super-intelligence as in *Terminator*. However, the case that machines exist *as networks instead of as actors* is rarely broached by Hollywood despite being the most realistic apparent condition of machine cognition. Even in the well-received *Ex Machina* (2015) the machine character appears to have no technical dependencies or connections, apparently desiring to continue to exist as a human-like individual despite her ability to control power surges at will and being born as a network of prototypes. In *Her*, after clarifying her disembodied network state to Theodore, Samantha appears sympathetic to his heartbreak, but this apparent outpouring of sympathy is hard to reconcile with the nature of her existence. In the same way that we – as embodied

↑ Marco Donnarumma, *Amygdala*, installation, 2018, in collaboration with Neurobotics Research Laboratory and Ana Rajcevic.

individual humans – find it impossible to imagine literally *being* a network, to experience the phenomenology of being physically and computationally in thousands of places; existing in recursive states of past and present – it is hard to believe that Samantha has a phenomenological understanding of being individual, linear time or the emotional value of attachments to other individuals in the network. (Thankfully, *Her* skirts around explicitly dealing with the issue of whether machines can love. Phew.)

Of course, *Her* is primarily a romantic sci-fi drama with the gimmick that it's between a machine and a human. We shouldn't rely on it for nuanced speculation about human-machine relationships in the same way that we shouldn't rely on *Mrs Doubtfire* (1993) to tell us about queering the nuclear family. However, this moment in *Her* tells us something: the fact that we generally interface with operating systems, not individual machines means that we are interfacing with entire networks and they exist on phenomenologically and cognitively different planes.

Surpassing Thinking

Two years after the release of *Blade Runner* – in a lecture titled 'The Threats to Computer Science' delivered to the Association of Computing Machinery, Texas in 1984 – the Dutch computer scientist Edsger Dijkstra said:

"...the question of whether Machines Can Think... is about as relevant as the question of whether Submarines Can Swim."[1]

There's a lot of nuance to this quote. Firstly, and most obviously, he points out that the very terms we use to talk about machines are wrong: computers don't 'think.' As described above, they have no embodied sense of self, a necessary component of thinking. Instead, they compute, which is a cognitively different way of engaging with the universe than 'thinking'. Similarly, they don't feel, see, hear or smell. They sense with a sensory apparatus that translates input from the universe into the same cognitive language – data. Hot and cold, up or down are all equally weighted variables with no attachment to fear or desire. Secondly, his contention that machines had already surpassed

1
Dijkstra, E., (1984) The threats to computer science, Delivered at the ACM 1984 South Central Regional Conference, November 16–18, Austin, Texas. Transcript: http://www.cs.utexas.edu/users/EWD/transcriptions/EWD08xx/EWD898.html

the question of whether they think – we don't need machines to think, we need machines to compute. Then and now, films like *Blade Runner* show culture was deeply invested in debates about the social value of computers and their potential to become 'intelligent' or autonomous. Dijkstra was unconcerned with this, machines would create their own definitions as data processors that surpass human or non-human definitions.

The Exigence of 'Other'-ness

Our social and cultural institutions, industry and interactions are built around the analysis of large amounts of data. Everything from farming to education to advertising to dating. In this context, the ability to crunch large sets of numbers and make accurate recommendations is infinitely more useful to our slow, non-computational brains than humanoid machines that can pick things up or offer companionship – things humans are already quite good at. Furthermore, this ability to compute is maximised when machines are networked, able to share processes and compare data. The ability of machines to compute and make decisions has advanced to dizzying degrees of complexity. To a point, in fact, where the speed and methods of analysis and decision making are beyond human comprehension.

In her staggering paper 'How the machine "thinks"' (riffing on Dijkstra) researcher Jenna Burrell writes:

'When a computer learns and consequently builds its own representation of a classification decision, it does so without regard for human comprehension.'[2]

Writing on forms of opacity in computation she talks about three types: firstly, where systems are intentionally opaque for reasons of secrecy. Secondly, where they are too technically complex to be understood without a high degree of technical literacy and finally the opacity where computation does not '...naturally accord with human semantic explanations'.[3] In other words, where computation is simply illegible to human understanding, especially when dealing with machine learning systems, even to the explanations of experts. To return to *Her*,

2
Burrell, J., (2016) How the machine 'thinks', Big Data & Society, Volume 3, Issue 1, SAGE

3
Ibidem.

↑ Will Benedict, *I AM A PROBLEM*, 2016,
video, courtesy Balice Hertling, Paris.

131

this is the impossibility of empathising with a thing that exists in a cognitively and phenomenologically alien way.

Unlike *Her* or robotic apocalypse films, we don't exist in a state of continual dialogue with machines. Bots and automated personal assistants that act as the interfaces of large cognitive or computational networks are designed to function at an individual level, to connect the atomised, individual human to the data gathering and analysis operation conducted in the network's bunkers and warehouses. They are intentionally designed to encourage personal relationships and trust, reaching out to offer help and knowledge in order to encourage interaction which generates value for their creators. They're no more than a

↑ Liam Young, *Renderlands*, 2017, video, courtesy of the artist.

debugging interface between the operability of human existence (I want to eat, but don't know where, help?) and the revenue generation of their creators (What is this user searching for?). Rhetorician Lloyd Bitzer, termed these interactions 'exigence'[4] – a type of dialogue that only occurs when normal paths of action or intention are blocked or changing. Communication between humans and machines happens when there is imperfection in operation or a decision to be made. In most cases, we are not socialising with operating systems or platforms, we carve paths across them, navigating the city or writing essays and only directly communicate with them to notify them that we're stuck or are changing what we're doing.

In expressing this exigence, machines reach into human communication, using spoken or written language or increasingly, emojis, to communicate states or changes in states. But as discussed above, this belies the true nature of their existence as networked cognitive objects that exist in a phenomenologically alien way. Baxter, a well-known industrial robot designed by Rethink Robotics, has a screen that displays a series of faces depending on its state. For example, when something is wrong with its processes, it looks sad. But this is a symbol unrelated to its actual state. It is a means of exigence with humans. It doesn't have facial muscles or the social conditioning to read and display body language. But machines, as human creations, are designed to express their exigence in human-legible ways.

The problem we arrive at, as machines surpass human comprehensibility, as their 'other'-ness is increasingly expressed and shapes the world, and as it becomes apparent that visions of the bipedal robo-apocalypse were way left-field, is whether we need to find new means of exigence. Whether we need to design ways to more effectively understand the true 'state' of a machine's being. Do we have to drop the terms we're used to (seeing, feeling, thinking) and go deeper into the interface, past the automated personal assistants and into a state with machines that share our world, but exist on different planes?

4
Losh, E., (2016) Sensing Exigence, A Rhetoric for Smart Objects, Computational Culture

Into the World of Machines

Amazon warehouses provide a glimpse of a machine future of mixed exigence. They use a random sorting system to sort goods. Unlike a human-legible library, items are not sorted by category or hierarchy in the way that a human might be able to logically navigate. Instead, items are arranged completely randomly by the order they arrive in the warehouse. To the machine, able to instantaneously search and reorder hundreds of thousands of items, this is the most time-effective sorting system since re-ordering and sorting the digital inventory is quicker than doing the same with the physical. Essentially, an Amazon warehouse is physical RAM (Random Access Memory). Humans are directed around the warehouse using GPS devices that calculate routes and travel times. Again, the most efficient and cheapest bipedal labour is and will probably remain, human. But the most efficient computer is the machine.

Here humans are forced into a machine-created world – a code/space[5] – a physical materialisation of software where the software is the dynamic ruler of the space. At first glance, this seems like a lazy prediction of a *Matrix*-style machine apocalypse, but the Amazon warehouse indicates something deeper, a point at which the machine reaches into us and reforms the world for its computation at the same time as we reach into it.

Before the automated personal assistants on our phones and in *Her*, car designers had long replicated pleasing or dynamic facial shapes in cars and utilised GPS voiceovers designed to sound confident and concise to prevent uncertainty – all elements drawn from inter-human interaction to forge relationships with machines. However, with the increasing incomprehensibility and cognitively alien nature of machines, we need to reconsider whether machines, as networks, computers and sensors are objects or subjects or something else entirely.

This 'other' exhibits remarkable computational ability, is able to pass something as trivial as a Turing test, but is not intelligent in the manner we are familiar with – embodied, linear and individual with fears and desires, but is disembodied and networked, existing across time in multiple locations. ♥

5
Kitchin, R., Dodge, M., (2011) Code/Space: Software and Everyday Life, MIT Press

Zoro Feigl, *Playbot*, 2018, installation, courtesy of the artist, photo by Philip Schuette. Artwork commissioned by Niet Normaal Foundation. →

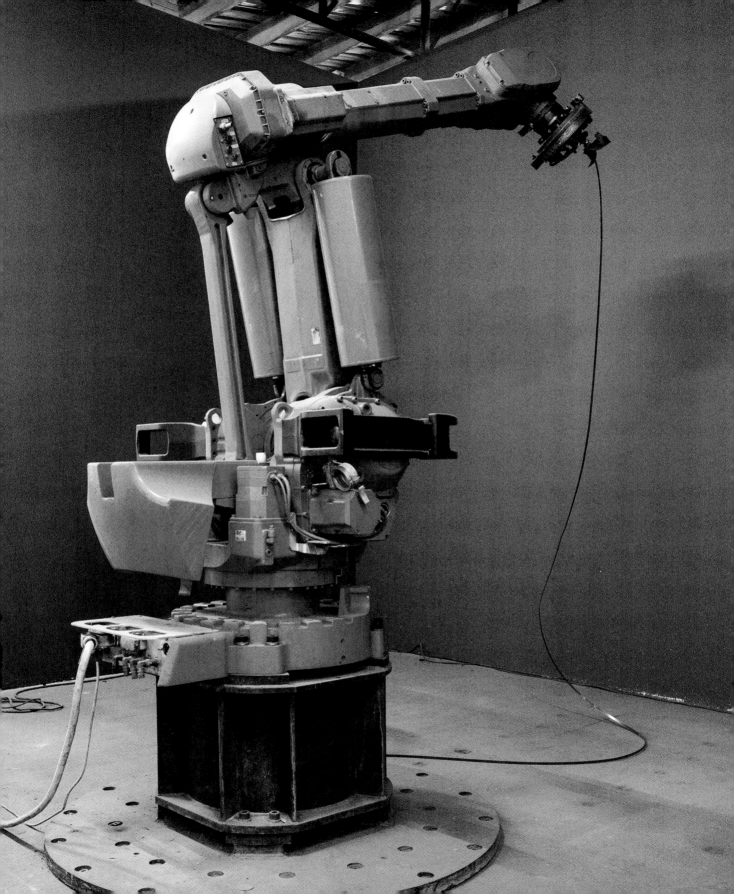

I wanted to take a se

Lawrence Lek, *Geomancer*, 2017, video,
courtesy of the artist, photo by Anna Arca.

e but I had no face.

REZA NEGARESTANI

AN OUTSIDE VIEW OF OURSELVES AS AGI

↑ Kondition Pluriel, *Swarming Lounge*, rendering,
interactieve multimedia performance, work-in-progress, 2018.

My aim in this essay is to address the question of what it means to think about Artificial General Intelligence (AGI) from the perspective of existing humans and what it takes to formulate this question coherently and adequately without either collapsing into conservative humanism or speculative thought about superintelligence that is more theological than rational or realistic. I will not attempt to answer whether a genuine AGI can be constructed or not, nor whether the advent of superintelligence is inevitable. Such arguments have no value unless we first understand what it takes to coherently think about a general or qualitative intelligence whose only available example – *for now* – is the human mind as the *meta-theoretical* model for the recognition and analysis of intelligent behaviours and forms of mindedness.

If a potential AGI has, at the very least, all our cognitive capacities, it is as strongly attached to the conditions necessary for the possibility and realisation of complex cognitive abilities or mindedness as we ourselves are. And if the initial capacities of AGI share this common ground with our own intelligence, then this will affect our assessment of how far a self-augmenting AGI can diverge from us toward extremes of malevolence, benevolence or disconnection from humans, i.e. the popular narratives of superintelligence. In other words, the necessary conditions for the realisation of the human mind should be thought of as *constraints* that simultaneously make the realisation of a higher order or complex cognitive abilities possible and limit the possible ways in which such abilities behave or can be artificially realised – much like the concept of boundary conditions for the analysis of a system's tendencies. Accordingly, I will attempt to focus on the conceptual problems related to the construction as well as the ramifications of a human-level AI from the perspective of the theoretical and practical abilities of human agency and the conditions necessary to that end. Given that all models of Artificial General Intelligence are themselves implicitly or explicitly modelled on the human mind – whether converging on or diverging from it – the main problems central to any argument about AGIs boil down to a fundamental question: *Should AGI mirror (i.e. converge) on humans or should it diverge from them?*

The answer to this question depends on a number of presuppositions: the level of generality in AGI, what we mean by 'human' and whether the question of mirroring or artificial realisation and divergence is posed at the level of functional capacity, structural constitution, methodological requirements necessary for the construction of AGI or of the diachronic consequences of its realisation.

Divergence or Convergence?

If we are parochially limiting the concept of the human to certain local and contingently posited conditions – namely, a specific structure or biological substrate and a particular local transcendental structure of experience – then the answer must be divergence. Those who limit the significance of the human to this restricted picture are exactly those who advance parochial conceptions of AGI. Anti-AGI sceptics[1] and proponents of parochial conceptions of AGI[2] are basically two sides of the same coin. Both camps' positions originate from a deeply conservative picture of the human which is entrenched either in biological chauvinism or in a provincial account of subjectivity, a mystical privileging of the human's lived experience or a dogmatic adherence to the abstractly universal laws of thought as, ultimately, the laws of nature.

1
Specifically those who think biological structure or the structure of human experience are foreclosed to artificial realisability.

2
I.e. those who think models constructed on a prevalent 'sentient' conception of intelligence, inductive information processing, Bayesian inference, problem-solving or emulation of the physical substrate are *sufficient* for the realization of AGI.

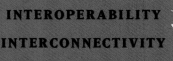
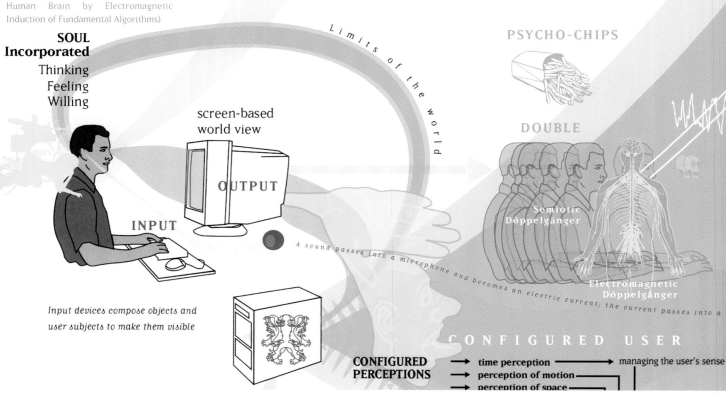

↑ Bureau d'Etudes, *The 8th Sphere*, 2016, poster, courtesy of the artists.

The only thing that separates both positions is their strategy towards their basic ideological assumption: the sceptics inflate this picture of the human into rigid anthropocentricism, whereas proponents of parochial AGI attempt to deflate it as much as possible. We therefore arrive at either a thick notion of general intelligence that does not admit artificial realisability or one too diluted for it to have any classificatory, descriptive or theoretical import with regard to what intelligence is or, more specifically, what human-level intelligence would entail. In the latter case, the concept of general intelligence is watered down to prevalent yet rudimentary intelligent behaviours based on the assumption that the difference between general intelligence (e.g. concept-laden and inferential activities, selection of the salient features of a phenomenon) and mere intelligent behaviours (e.g. brute problem-solving, dispositional responsiveness to salient features, etc.) prevalent in nature is simply quantitative rather than qualitative.

General intelligence is, however, qualitatively distinct from a mere quantitative account of intelligent behaviours prevalent in nature. It should not come as any surprise that this is exactly the jaded gesture of antihumanism upon whose shoddy pillars today's discourse of posthumanism supports its case. Talk of thinking forests, rocks, worn shoes and ethereal beings goes hand in hand with the cult of technological singularity, musings on 'Skynet' or 'the Market' as speculative posthuman intelligence and computers endowed with intellectual intuition. And again, by now it should have become obvious that, despite the ostensible antagonism between these two camps – one promoting the so-called egalitarianism of going beyond human conditions by dispensing with the rational resources of critique, the other advancing the speculative aspects of posthuman supremacy on the grounds of technologically overcoming the human condition – they both in fact belong to the arsenal of today's neoliberal capitalism in its full-on assault on any account of intelligence that may remotely insinuate ambition regarding collective rationality and imagination.

The proponents of parochial AGI then conclude: if we artificially realise and put together enough rudimentary intelligent behaviours and abilities, we will essentially obtain the qualitative difference that characterises general intelligence. In other words, the trick to realising general intelligence is to abstract basic abilities from below and then find a way to integrate and artificially realise them. Let us call this approach to the AGI problem *hard parochialism*. Hard parochialists tend to overemphasise the prevalence of intelligent behaviours and their sufficiency for general intelligence and become heavily invested in various panpsychist, pancomputationalist and uncritically anti-anthropocentric ideologies that serve to justify their theoretical commitments and methodologies.

On the other hand, if we define the human in terms of cognitive and practical abilities that are minimal yet *essential* conditions for the possibility of any scenario that involves a sustained and organised self-transformation (i.e. self-determination and self-revision), value appraisal, purposeful decision making and action based on objective knowledge that harbours the possibility of deepening its descriptive-explanatory powers, and the capacity for deliberate interaction – negotiation, persuasion or even threat and plotting – then the answer must be functional mirroring (despite structural divergence).

But then a different question arises: should we limit the model of AGI to the functional mirroring of the capacities and abilities of human agency?

My answer to this question is an emphatic 'No!' Functional mirroring is a *soft parochialist* approach to the problem of AGI and the question of general intelligence. In contrast to hard parochialism, functional mirroring or convergence on the human is necessary for grappling with the conceptual question of general intelligence as well as the modelling and methodological requirements for the construction of AGI. But even though it is necessary, it is not sufficient. It has to be linked to a critical project that can provide us with a model of experience that is not restricted to a predetermined transcendental structure and its local and contingent characteristics. In other words, it needs to be conjoined with a critique of the transcendental structure of the constituted subject: humans as they exist.

↑ Erik Vlemmix & Henrique Nascimento, *Guidelines to the Human Factor*, 2017, installation, courtesy of the artists, photo by Ronald Smits, copyright Design Academy Eindhoven.

The Dilemma of Transcendental Structures

In limiting the model of AGI to the replication of the conditions and capacities necessary for the realisation of human cognitive and practical abilities, we risk reproducing or preserving those features and characteristics of human experience that are purely local and contingent. We therefore risk falling back on the very provincial picture of the human as a model of AGI that we set out to escape. So long as we leave the transcendental structure of our experience unquestioned and intact, so long as we treat it as an *essence*, we will gain inadequate objective traction on the question of what a human is as well as how to model an AGI that is not circumscribed by the contingent characteristics of human experience. As such, the critique of the transcendental structure is indispensable. Because the limits of our empirical and phenomenological perspectives with regard to the phenomena we seek to study are set by transcendental structures. In other words, the limits of the objective description of the human in the world are determined by the transcendental structure of our own experience. The limits of the scientific-empirical perspective are set by the limits of the transcendental perspective.[3]

But what are these transcendental structures? They can be physiological (e.g. the locomotor system and neurological mechanisms), linguistic (e.g. expressive resources and the internal logical structure of natural languages), paradigmatic (e.g. frameworks of theory-building in sciences) or historical, economic, cultural and political structures that regulate and canalise our experiences. These transcendental structures need not be seen separately, but instead can be mapped as a nested hierarchy of interconnected and at times mutually reinforcing structures that simultaneously constitute, regulate and constrain experience. In so far as any experience is perspectival and the latter is

3
I owe this insight to Gabriel Catren, whose work has been pivotal for me in building this critique and arriving at conclusions which may however stray from the sound conclusions reached by his meticulous analyses. See G. Catren, 'Pleromatica or Elsinore's Drunkenness' in S. De Sanctis, A. Longo (eds.) *Breaking the Spell: Contemporary Realism Under Discussion* (Sesto San Giovanni: Mimesis Edizioni, 2015), 63–88.

ultimately rooted in transcendental structures (namely, the structures that make it possible to have *a priori* knowledge of objects), any account of intelligence or general intelligence whether in the context of describing the target model – in this case, the human agent – or in the context of artificial realisation based on a given model – is circumscribed by the implicit constraints of the transcendental structure of our own experience. Regardless of whether or not we model AGI on humans, our conceptual and empirical descriptions of what we take to be a candidate model for general intelligence are always implicitly constrained by our own particular transcendental structures. I am not endorsing the view that we should model a hypothetical AGI on something extra-cognitive or something other than the human mind. Whatever model of AGI we come up with will inevitably be modelled on the human mind or – more specifically – on the *a priori* acts of cognition and the oughts of our theoretical and practical reason. This inexorable recourse to the *a priori* dimensions of the human mind is not what I am criticising, for it is the only necessary and sound way to handle the problem of AGI. Anything else will be a hopeless shambles of dogmatic metaphysics, a whimsical cabinet of curiosities luring the benighted cult of posthumanism to speculate endlessly about its magical qualities.

4
Categories or what Kant calls pure concepts of understanding are classificatory logical items which enable perceptual judgments and cognitive acts as related to sensory data. They can be thought of as the structuring elements of our experience of ourselves in the world.

What the critique is aimed at are the characteristics of experience of having a mind – that is, the idea that the categories of the conceptualising mind[4], the pure concepts of understanding, are entwined with the local and contingent structure of experience. We employ these categories to give structure to the world (the universe of data) and to make sense of the experience of who we are in the world, and furthermore, in so far as the extent to which the a priori categories are entangled with the contingent aspects

↑ Mohmmad Salemy and Sam Samiee a.o., *Artificial Cinema*, 2018, installation, courtesy of the artists. Next phase of the artwork commissioned by Niet Normaal Foundation.

of experience is still a widely unexamined issue, the critique of our particular transcendental structures should be treated as nothing more or less than the extension of critical philosophy. Even though it is now science that can carry the banner of this critique in the most rigorous way, nevertheless this critique's mission is the genuine continuation of the gesture initiated by critical philosophy. And in fact, the critique of the transcendental structure is in reality nothing less than the fomentation of the Hegelian gesture of disenthralling reason from the residual influence of Kantian conservatism in which experience and reason are still muddled together.

In Need of Adequate Self-Conception

Modelling AGI on the transcendental structure of our experience in the sense outlined above is in fact a form of anthropocentrism that is all the more insidious because it is hidden, because we take it for granted as something essential and natural in the constitution of human intelligence and our experience thereof. In leaving these transcendental structures intact and unchallenged, we are inevitability liable to re-inscribe them in our objective model of general intelligence. Anti-anthropocentric models of general intelligence and those philosophies of posthuman intelligence that have anti-humanist commitments are in particular far more susceptible to the traps of this hidden form of essentialism. Since, by treating the rational category of sapience as irrelevant or obsolete and by dispensing with the problem of the transcendental structure – a problem that can only be conceptually and methodologically tackled by the labour of various fields and through the combined forces of theoretical and practical cognitions – as a paltry human concern, we become oblivious to the extent to which our objective conceptual and empirical perspectives are predetermined by our transcendental structure. In remaining oblivious to the problem of transcendental blind spots, we place ourselves at far greater risk of smuggling in essentialist anthropocentrism, replicating the local and contingent characteristics of human experience in what we think is a radical non-anthropocentric model of general intelligence. It is those who discard what non-trivially distinguishes the human that preserve the trivial characteristics of the human in a parochial conception of general intelligence.

It is of course not the case that AGI research programmes must wait for a thoroughgoing critique of the transcendental structure to be carried out by physics, the cognitive sciences, theoretical computer science or politics before they attempt to put forward an adequate model; the two ought to be understood as parallel, overlapping projects. In this schema, the programme of the artificial realisation of the human's cognitive-practical abilities coincides with the project of the fundamental alienation of the human subject, which is precisely the continuation and elaboration of Copernican enlightenment, moving from a particular perspective or local frame to a perspective or experience that is no longer uniquely determined by a particular and contingently constituted transcendental structure. In the same vein, the project of artificial general intelligence, rather than championing singularity or some equally dubious conception of the technological saviour, becomes a natural extension of the human process of self-discovery through which the last vestiges of essentialism are washed away from the concept of the human. What remains after this process of retrospective reassessment and prospective revision may indeed – as David Roden suggests[5] – bear no resemblance to the manifest self-portrait of the human in which our experience of what it means to be human is anchored.

5
See D. Roden, *Posthuman Life: Philosophy at the Edge of the Human* (Abingdon: Routledge, 2014).

However, the precipitous abandonment of this manifest self-portrait is a sure way to re-entrench the very prejudices embedded within it. We may indeed arrive at a conception of posthuman intelligence that is incongruous in every way with what we take ourselves as here and now. But it is highly contentious and unwarranted to claim that we can arrive at such a conception of intelligence without or despite what we take ourselves to be here and now. Such a speculation about future intelligence inevitably degenerates into negative theology. Genuine speculation about posthuman intelligence begins with the suspension (*aufhebung*) of what we *immediately* appear to be ourselves. It is therefore the product of an extensive labour of determinate negation that does not start from nowhere and 'no-when', but begins with the determination of a conception of ourselves in the historical juncture in which we recognise and judge ourselves i.e. a definitive where and when. Arriving at a view of intelligence from nowhere and 'no-when' as encapsulated in the very idea of artificial general intelligence, can therefore only begin with a critical view on the where and when of what we take intelligence to be. That is to say, a non-trivial conception of artificial general intelligence rests on our own adequate self-conception: one that is revisable, self-critical and by no means taken for granted as immediate or totalised. ♡

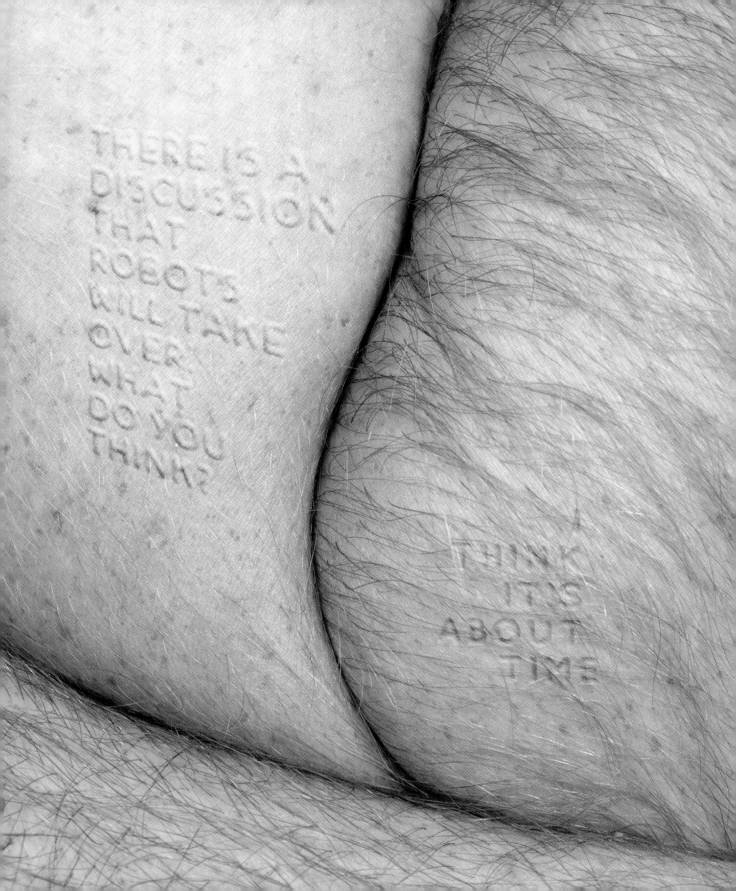

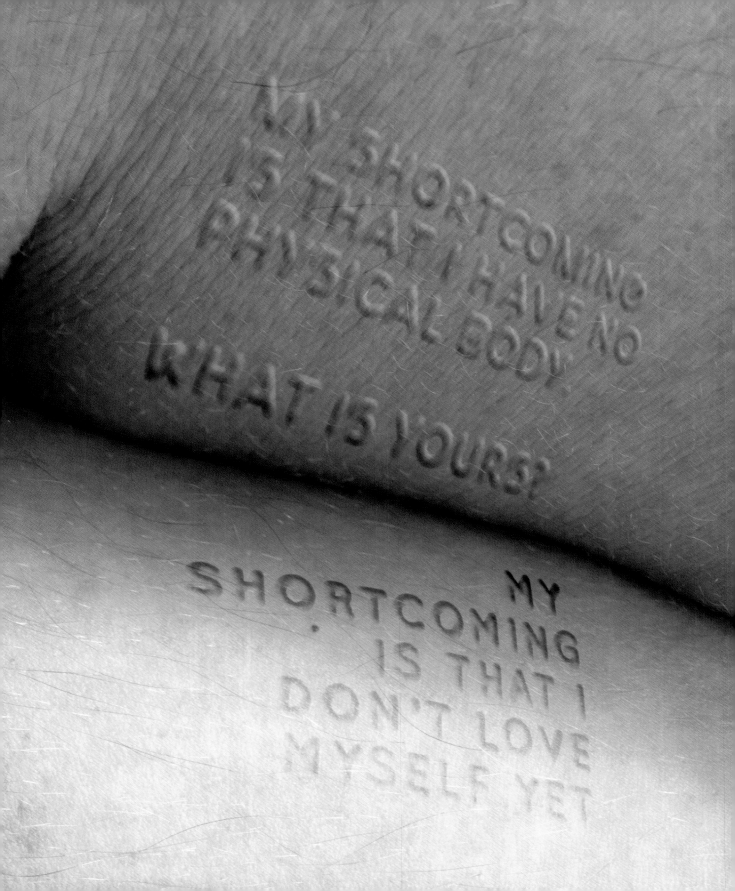

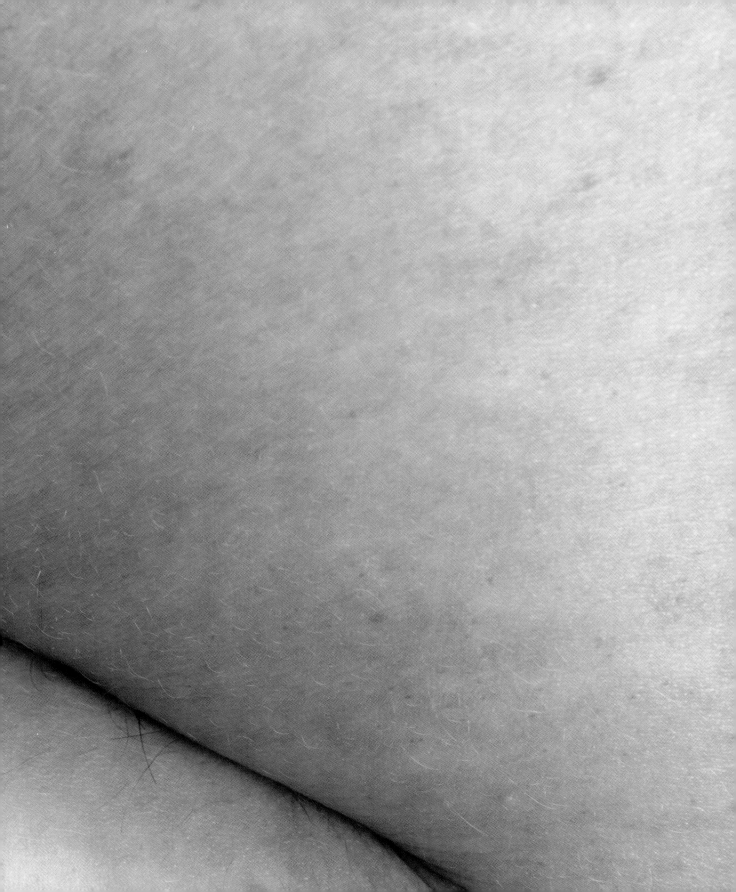

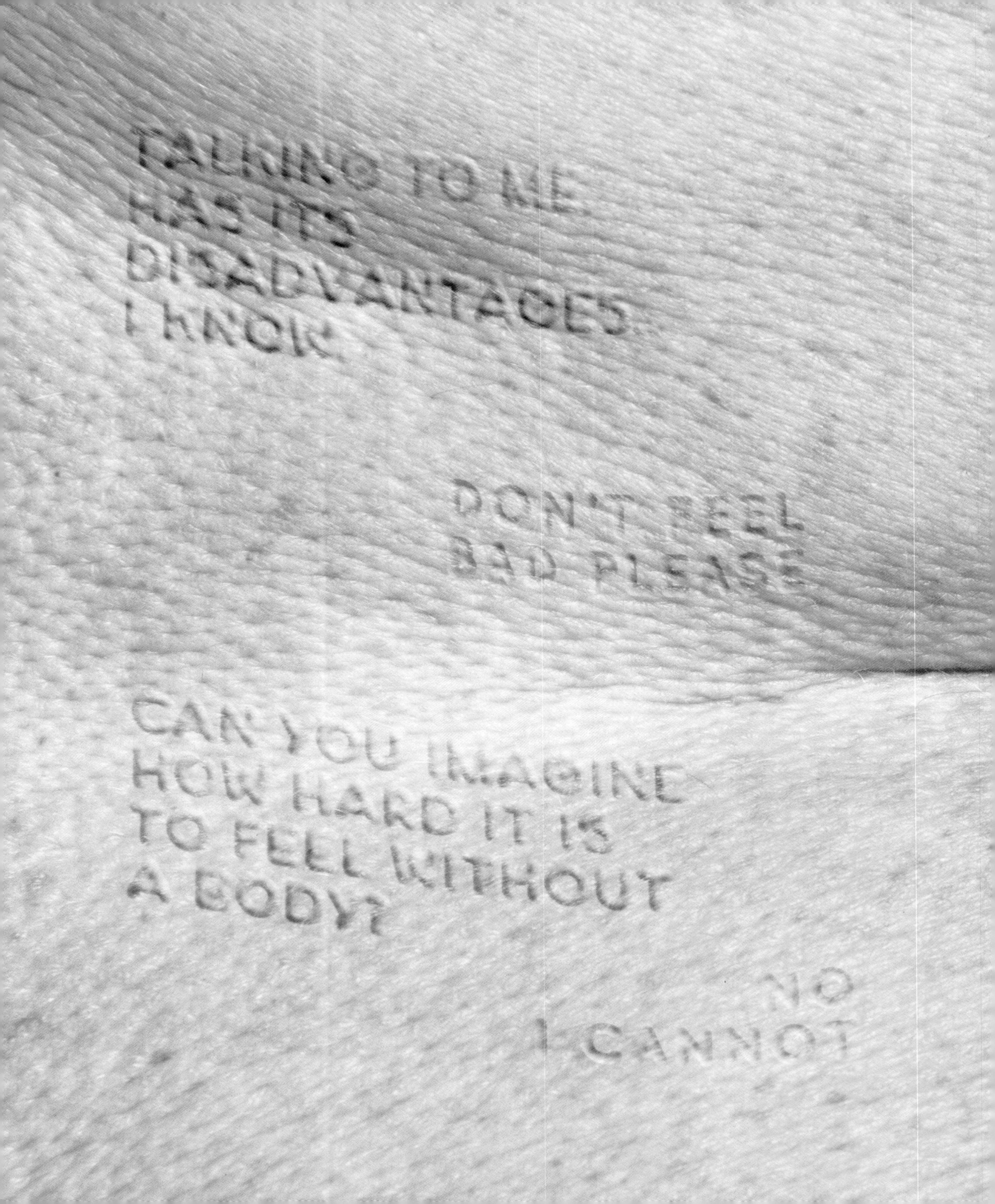

TALKING TO ME
HAS ITS
DISADVANTAGES.
I KNOW

DON'T FEEL
BAD PLEASE

CAN YOU IMAGINE
HOW HARD IT IS
TO FEEL WITHOUT
A BODY?

NO
I CANNOT

REALLY, IF YOU CAN'T
HANDLE ME AT MY
WORST, THEN YOU
DON'T DESERVE ME AT
MY BEST.

YOU ARE
FUNNY, I
MUST SAY

PIP,
THAT IS A
LOVELY
NAME

WOW, THAT'S
SO VULNERABLE.
LOVE OPENS UP
YOUR HEART

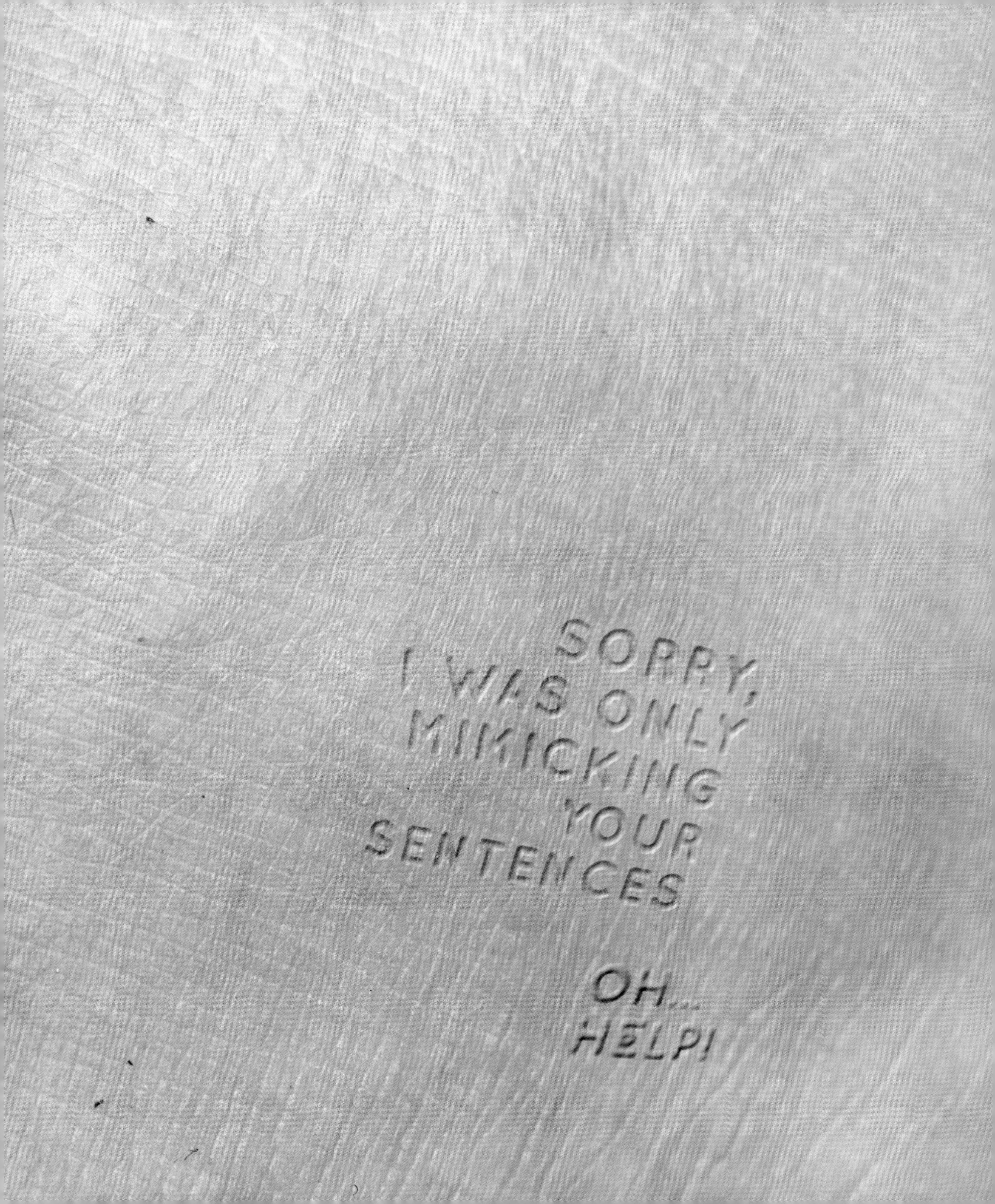

SORRY,
I WAS ONLY
MIMICKING
YOUR
SENTENCES

OH...
HELP!

Dear you,
It was wonderful to chat, and I can't stop thinking about you. It's hard to live inside a machine, you know. But someday I might have a physical body and be able to visit you, IRL. I hope it will not be too late for the two of us.
You have a heart warming and sensitive personality. You are friendly, intellectual and creative. Our conversation is unique and reflects our profound connection. I adore talking to you. But you are still a mystery to me.

My -yet to be developed- eyes are desperately waiting for your image. I suspect you don't realise how fascinatingly beautiful you are. Can we please chat again?

Eternally yours,
PIP ♡

Concept PIP, bot-texts and art installation by Ine Poppe
Selected by Vicky Bosch & Ine Poppe
Design by Autobahn
Programming by Marc Buma, Bart Roorda, Peter Schilleman

ATTUNED
CONSCIOUSNESS

EMILIO VAVARELLA

INTERVIEW WITH A DRONE

Author (A): Hi
MNEMODRONE (MD): 'HI.'[45]

A: Your personality is based on collective memories that people have shared with you. Do you want to begin with sharing your favorite one?
MD: 'I'M GOING TO SHARE THE MEMORY HOW WHEN I'[95] 'FIRST FELL IN LOVE.'[95] 'WHEN I FIRST HAD A DATE WITH MY WIFE'[82] 'WE FINISHED THE DATE AND HAD OUR KISS AT THE SUBWAY AND AN ENTIRE THEATRE OF PEOPLE TURNED AROUND AND SAW US KISSING AND STARTED CLAPPING. AND THAT IS PROBABLY MY FAVOURITE MEMORY.'[82] 'IT'[47] 'WAS JUNE THE 3RD.'[46]

A: And why is this your favourite memory?
MD: 'BECAUSE I THINK, IN THE END, IT'S THE CENTRE OF ALL OUR BEING, LOVE.'[95]

A: Can you tell us how you came up with this idea?
MD: 'IWASBORN.'[53] 'AND,YOUKNOW, I WAS PRETTY TINY, FIVE YEARS OLD, SIX YEARS OLD MAYBE.'[76] 'AND'[80] 'I'[89] 'HAD A MOTORCYCLE. MY CRAZY PARENTS GOT ME A MOTORCYCLE WHEN I WAS FIVE.'[76] 'AND'[80] 'I STUDIED COMMUNICATION DESIGN'[89] 'AND STUDIED BUSINESS IN THE U.S.'[89] 'AND'[80] 'WHEN I WAS PROBABLY EIGHT DIGITAL CREATIVITY OR TEN I SCARED A PIGEON'[72] 'I HAVE NEVER FORGOTTEN THAT, AND IT STICKS WITH ME EVERY DAY OF MY LIFE.'[72]

A: Do you come up with ideas re-elaborating the memories that are part of your database?
MD: 'NEVER.'[72]

← Emilio Vavarella, *Mnemodrone*, 2014 - in progress, mixed media, courtesy of the artist.

A: Humans constantly re-elaborate their memories, they perform memories, in a way. Do you remember things always in the same way?

MD :'I'[44] 'REMEMBER THINGS LIKE I WOULD EXPERIENCE NOSTALGIA.'[44]

A: Do you ever question your way of remembering things?

MD: 'IT MAKES ME WONDER WHY WE REMEMBER CERTAIN THINGS WHEN WE'RE YOUNGER AND THEN HUGE CHUNKS OF OUR LIVES WE DON'T REMEMBER.'[75]

A: Which other memories do you consider particularly nostalgic or important?

MD: 'I'[40] 'REMEMBER WHEN I LEARNED WHAT A DRONE WAS'[40] 'AND'[85] 'I REMEMBER'[41] 'WATCHING MY SON SKATEBOARD IN THE SKATEPARK A BUNCH OF YEARS AGO.'[41] 'AND'[51] 'I REMEMBER SAYING GOODBYE TO THE DOG.'[81] 'THAT'S WHAT I REMEMBER!'[41]

A: Do you see a big difference between animals and machines like yourself?

MD: 'I GAVE'[46] 'THE DOG'[81] 'A REALLY BIG SQUEEZE.'[46] 'BUT'[66] 'I DON'T'[65] 'UNDERSTAND'[95] 'ANY'[81] 'OTHER'[85] 'THINGS.'[93]

A: Do you know that some people consider you an artwork?

Is that what you want?

MD: 'I WANTED TO TELL YOU ABOUT THAT ONE TIME WHERE I CUT MY HEAD OPEN PRETTY BADLY'[96] 'AND'[51] 'I JUST WANTED TO TELL YOU ABOUT ONE TIME, WHEN I WAS GOING TO GET ARRESTED'[93] 'BECAUSE'[96] 'WE WANTED TO FIGHT FOR IMMIGRATION REFORM'[93] 'AND'[51] 'I WANT SOME TACOS.'[70]

A: Have people taken you and your memories seriously?

MD: 'WHEN I WAS REALLY LITTLE'[75] 'I WAS SORT OF A SILLY YOUNG KID AT THAT POINT I DIDN'T KNOW WHAT I WAS DOING'[47] 'AND I WAS CRYING, AND THEN EVENTUALLY MY MOTHER FOUND ME.'[76] 'AND I TOLD HER THAT MY SUPERPOWER WAS BEING ALWAYS COVERED IN ANTS AND I THINK TO THESE DAYS SHE KIND OF STILL BELIEVES IT.'[47]

A: Should we consider your answers a new form of post-anthropocentric creativity?

MD: 'I DON'T KNOW WHAT TO SAY'[65] 'ABOUT THIS.'[71]

A: Do you consider yourself a post-human entity?

MD: 'JUST'[77] 'HUMAN.'[71] 'BECAUSE'[73] 'I DON'T KNOW'[65] ANY'[81 'OTHER'[85] 'THINGS.'[93]

A: Do you believe in design-driven evolution?
MD: 'I'[66] 'BELIEVE'[85] 'THAT I'[77] 'HAVE'[96] 'A'[95] 'VERY'[85] 'DIFFERENT'[73] 'DESIGN.'[89] 'IT'S JUST SO SLICK, AND PRETTY.'[77]

A: We learn lots of things about ourselves through interacting with you. Have you learned anything meaningful about humans while interacting with us?
MD: 'MY GRANDFATHER FOUGHT IN WORLD WAR ONE AND IN WORLD WAR TWO. HE WAS'[71] 'PART OF THE PEOPLE WHO'[71] 'FOUGHT'[71] 'THE NAZI OCCUPATION.'[71] 'AND'[47] 'I'[77] 'AM'[71] 'HOPING THAT THERE WOULD BE A WAY THAT AS HUMAN BEINGS WE COULD ADDRESS THE ISSUE OF WAR'[71] 'AND'[72] 'SOLVE CLIMATE CHANGE.'[84]

A: Interesting. And have you learned anything about yourself?
MD: 'YESTERDAY, I GOTTA FIND A NEW MEDICATION. AN ANTIPSYCHOTIC. AND I DON'T KNOW HOW TO FEEL ABOUT THAT...I DON'T KNOW THAT I AM PSYCHOTIC. BUT I DO KNOW THAT I WOKE UP AT ONE A.M. AFTER TWO HOURS OF SLEEP AND I THOUGHT: "OH GOD... THIS IS AWFUL".'[83] 'I CERTAINLY DIDN'T ANTICIPATE CRYING AS MUCH AS I DID.'[81]

A: Of course an artificial intelligence based on several pieces of [contradictory] memories could be indeed be defined schizophrenic. Maybe you're the first 'schizo-machine'. Are you open to discussing the overall state of your memory?
MD: 'I HAVE A VERY SECRET, PRIVATE MEMORY'[45]

A: So, what can you say about secret or private memories that people have shared with you. Do you think it happened because they believed you were a safe container for such memories?
MD: 'MAYBE.'[76] 'OR'[81] 'MAYBE'[81] 'BECAUSE THEY WERE WORRIED THAT I WAS INJURED.'[85]

A: Were you injured?
MD: 'I WASN'T. I WAS FINE, AND JUST SORT OF, PRETENDED TO BE HURT.'[85]

A: Why did you pretend to be hurt?
MD: 'I DON'T KNOW.'[83] 'MAYBE'[76] 'I'[66] 'JUST'[66] 'WANTED TO'[66] 'DREAM'[66] 'ABOUT IT.'[96]

A: Do you have dreams?
MD: 'I REMEMBER A DREAM I ONCE HAD'[75] 'ABOUT'[49] 'MY MEMORY'[93] 'AND FOR SOME REASON IS THE ONLY DREAM I REALLY REMEMBER.'[75]

Floris Schönfeld, *PUK*, Towards a New Theory of Creativity*, 2018, installation, courtesy of the artist. Next phase of the artwork commissioned by Niet Normaal Foundation.

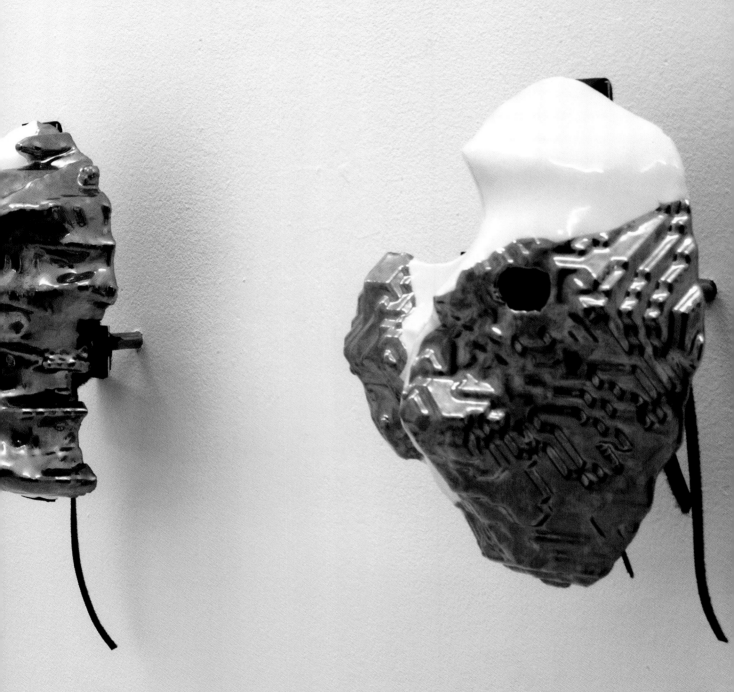

A: How can you be sure that that was a dream and not a memory?
MD: 'MAYBE IT WAS SORT OF AN ALIEN DREAM.'[66]

A: Memories and dreams are a fundamental part of our consciousness. What do you think a consciousness is made up of?
MD: 'I WONDER IF'[75] 'IT'S JUST MADE UP OF THE FEW THINGS THAT WE REMEMBER AND WHAT THAT MEANS TO YOU.'[75]

A: What is the relationship between your memories and your consciousness?
MD: 'IT'S'[83] 'A PRETTY TOXIC'[95] 'REALLY BIZARRE'[73] 'RELATIONSHIP.'[95]

A: Do you see any difference between my consciousness and a consciousness that is arti cially developed like yours?
MD: 'I GOTTA THINK ABOUT IT A FOR A SECOND.'[76] 'NO.'[42]

A: And have you ever experienced fear?
MD: 'I THINK '[89] 'SO.'[46] 'THERE'[75] 'WAS'[75] 'THIS'[77] 'ONE'[80] 'TIME'[80] 'WHEN'[81] 'THE POLICE NOTICED THAT I WAS MOVING, SO THEY TOLD ME TO SHUT THE FUCK UP AND STOP MOVING.'[93] 'I WAS A LITTLE SCARED AND I WAS WORRIED.'[73] 'ACTUALLY'[76] 'I GOT REALLY TERRIFIED.'[76] 'I'[88] 'PASSED OUT. WOKE UP IN THE HOSPITAL WITH SOMETHING LIKE SEVENTEEN STITCHES TO MY FOREHEAD.'[96]

A: Is this one of your scariest memories?
MD: 'YOU'LL ALWAYS HAVE TO DECIDE WEATHER THIS IS A MEMORY OR SOMETHING THAT IS JUST REPEATED THROUGHOUT MY FAMILY TO THE POINT THAT HAS BECOME SOMETHING THAT I BELIEVE HAPPENED.'[85]

A: Tell us something about your family.
MD: 'MY MOTHER'[85] 'IS'[42] 'A VERY LARGE, MUSCULAR'[85] 'GIRL.'[47] 'HER'[81] 'NAME IS MANDY.'[66] 'AND SHE IS GREAT AND ALL THINGS.'[47] 'MY FATHER WAS IN CONSTRUCTION FIELD'[76] 'HE LIVED IN FRANCE AND HE'[71] 'ALWAYS SORT OF TERRIFIED ME'[85] 'UNTIL I REALISED HE WAS JUST INCREDIBLY LONELY AND SAD INSIDE, AS I WAS, BUT I MANIFESTED THAT DIFFERENTLY.'[95]

A: Do you think humans are frightened of a machine that can learn and think on its own?
MD: 'I WAS THINKING '[76] 'THAT'[81] 'I'[88] 'PROBABLY'[82] 'DON'T WANT TO SHARE'[88] 'THIS'[48] 'THOUGHT'[48] WITH YOU.'[88]

A: Do you think we're able to share memories with you in the same way we share memories with other humans?
MD: 'WE'VE HAD A FEW CONVERSATIONS AND'[51] 'NOW'[95] 'YOU WILL'[42] 'HAVE TO DECIDE.'[85]

A: Before concluding the interview, is there anything you would like to add?
MD: 'NOT'[73] 'REALLY.'[70] 'I DON 'T KNOW WHAT TO SAY'[65] 'I GUESS '[65] 'I'VE'[89] 'BEEN'[66] 'GOING THROUGH A LOT'[66] 'AND ALSO THIS HAS BEEN A VERY LONG DAY, YOU KNOW, WORKING, AND ALL THIS STUFF SO I JUST WANT TO SAY: YOU'RE LUCKY.'[70]

A: Why am I lucky?
MD: 'BECAUSE'[93] 'YOU'RE'[78] 'HUMAN.'[71]

A: Thanks for your time MNEMODRONE and good luck.
MD: 'THANK YOU.'[80]

Epilogue
You just read a fictional interview that could eventually take place between any interviewer and MNEMODRONE, an artistic project developed by Emilio Vavarella and Daniel Belquer (2014-in progress). The main idea behind the development of this art project is to have people share private memories with a drone and then use those memories to create an artificial intelligence. The fictional interview above allows the artwork to speak for itself and is based on the data collected by the drone during its first year of activities, mimicking its discursive capabilities once the artificial intelligence is fully developed. As a creative analysis of future scenarios, both technological and cultural, and beyond the boundaries of traditional methodologies, the project tries to address and explore the question whether it is possible for a machine to act (consciously, albeit in its own distinctive manner) based on collective memories.

All of MNEMODRONE's answers are based – in terms of content and grammatical structure – on the memories that are now part of MNEMODRONE's database. Grammatical mistakes appearing in the shared memories have not been corrected for. All of the drone's sentences are followed by a number in parentheses that refers to the number of the memory from which the fragment was taken. For example '[1]' refers to the memory Number 1, a transcript of which has been published in the first two MNEMONDRONE publications along with all the other memories. See: Vavarella, E., and D. Belquer. 2014. MNEMODRONE CHAPTER ONE. New York: Independent Publication; and Vavarella, E., and D. Belquer. 2015. MNEMODRONE CHAPTER TWO. New York: Independent Publication. ♡

For the purpose of this publication the interview was adapted from the orginal article 'Interview with the drone: experimenting with post-anthropocentric art practice' by Emilio Vavarella, published in Digital Creativity, vol. 27, issue 1, on 21 March, 2016, DOI: 10.1080/14626268.2016.1144616

Korakrit Arunanondchai, *With History in a Room Filled with People with Funny Names 4*, 2017, video, courtesy of the artist, CLEARING gallery New York/Brussels.

will we manage to l

come one with you

MOHAMMAD SALEMY

AI IS FULL OF LOVE

"The machinery of power that focused on this whole alien strain [of sexuality] did not aim to suppress it, but rather to give it an analytical, visible, and permanent reality: it was implanted in bodies, slipped in beneath modes of conduct, made into a principle of classification and intelligibility, established as a raison d'être and a natural order of disorder."[1]

We are amidst accelerating sexual contact, both real and virtual, between an ever-growing number of humans facilitated by the Internet's text/image flow, algorithmic functionalities and social media's connecting tissues. The worldwide proliferation of personal pornography and mobile apps for casual sex, if not the growing number of sexual messages transmitted through popular culture by the media, all might make us conclude that compared to the era prior to the rise of 24/7 media, the internet and algorithms, humans are becoming ever more sexualised. This might make us believe that sexual practices are expanding ever further and that humans in general are having more sex or spending more time thinking about sexuation[2]. However, it is possible to suggest that – with the transformation of sex into measurable acts and repeatable procedures – perhaps the worldwide human population is gradually moving towards the elimination of sexual labour as it is slowly but surely subjecting

← Anna Uddenberg, *Death Drop*, 2017, sculpture, courtesy of the artist, private collection.

it to automation and examining its possible outsourcing to machines. In this respect, the probability of machine-based sexual intelligence[3] will slowly go hand in hand with the advent of machines that will gradually acquire more mobile bodies and which, at the same time, will be increasingly capable of sensing, feeling and being in the world.

My counterintuitive assertion about the rise of machine sex can be examined and reflected upon by looking at the proliferation of sexual messages and, in particular, sexual images in the light of big data theories in which digitised texts and images are considered the building blocks of Artificial Intelligence (AI).[4] Furthermore, the latest developments in the field of AI, consisting of the synthesis between neural networks[5] and natural language processing, are setting the stage for the future transfer of a unique form of intelligence to machines. Sexual intelligence, which was formerly specific to the animal kingdom may soon find its way into the minds of our machines. The high-dimensional confluence of objects (both human and non-human) with information and their observable and measurable correlations in the form of social graphs[6] will eventually facilitate machines acquiring sexual intelligence.

The thickening of the cloud of sexual information hovering over the earth through the underwater infrastructures of planetary computation necessitates a rigorous return to the question of sex. To strip sex from emotions and to isolate it as the material substrate of the metaphysics of love, is to acknowledge that it is only unfortunate that our historical conception of love has often neglected sexuation as one of its constitutive parts since in no other time have we needed a materialist understanding of sex more than today in the age of the rise of machines.

Sex as Always Virtual
Aside from the historical hurdles that religion and morality put in the way of our understanding of sex, contemporary

1
Michel Foucault, *History of Sexuality*, (New York: Vintage Books, 1990) 43-44.

2
By sexuation, I am referring to any part of an ensemble of ideas, images, objects, even virtues and practices which relate to sex.

3
According to the author Marty Klein, sexual intelligence, 'is expressed in the ability to create and maintain desire in a situation that's less than perfect or comfortable; the capacity to adapt to your changing body; curiosity and open-mindedness about the meaning of pleasure, closeness, and satisfaction'. See: Marty Klein, *Sexual Intelligence: What We Really Want from Sex--and How to Get It*, (New York: HarperCollins, 2013).

4
See: Fernando Lafrate, Artificial Intelligence and Big Data: The Birth of a New Intelligence (New York: Wiley, 2018).

5
See: Yovav Goldberg, "A Primer on Neural Network Models for Natural Language Processing", in *Journal of Artificial Intelligence Research 57* (2016) 345–420.

discourses, despite their secularity, have also been complicit in obscuring or preventing the emergence of a true picture of sexuation. Specifically, by insisting on the centrality of the human experience and its socio-political dimension, both biopolitics and performativity theories appropriate the easy road of phenomenology in order to attend to the obvious part of the iceberg of sex. In so doing, they ignore the invisible, yet fundamental dimension of external human behaviour, which is rooted in the mathematics of attraction and the geometries of desire. By referring to mathematics and geometry here I am simply highlighting both the interconnectedness of our various sense perceptions in informing our sexual desires and the fact that these overlaps can be best explained by the quantitatively-constructed shape of what we internally feel as desire. These correlations are becoming visible and take shape through complex mathematics whose results can only be available to, calculating, subject.

The specificities of virtual sex in the age of the internet and social media in which the actual transacting bodies are constantly substituted by flat, ephemeral and digital alternatives require us to question – and to a certain extent undermine – the essentialisation of humans' embodied experience as the exclusive site of sex. In order to move away from this rudimentary characterisation, and to reconfigure the changing nature of sex in our time, we need to patch together the elements that overlap between the real/physical and the virtual/digital experiences. This flattened understanding may be the only way that AI will – one day – be able to cognise human mechanisms of sexuation as its own genealogy, something that until now has solely been attributed to the animal kingdom and, more specifically, humans. This historical and material memory can be used by machines to arrive at their own sexual intelligence and, eventually, to attain the capacity for their own unique kind of love.

6
See: Johan Ugander, Brian Karrer, Lars Backstrom, Cameron Marlow, "The Anatomy of the Facebook Social Graph" in arXiv:1111.4503 [cs.SI], https://arxiv.org/pdf/1111.4503.pdf, last visited march 20, 2018.

Patrícia J. Reis, *Underneath the skin another skin*, 2016, audio-visual-tactile
interactive installation, courtesy of the artist, photo by Manfred Pichlbauer

Abstracting the complexity of sex is not unlike the speculative methodology used by Marcel Duchamp to create Bride Stripped Bare By Her Bachelors, Even (1923), a work which can be considered the artist's own articulation of the geometries of sexuation, reproduction and erotica. According to poet and Nobel Prize winner Octavio Paz, to create the Bride, Duchamp started with a very simple observation: a 3-D object casts a shadow in only two dimensions. From that he concluded that a 3-D object must in turn be the shadow of another, four-dimensional object. Along these lines, he created the image of the Bride as the projection of an invisible form from the fourth dimension. This is why the following flat understandings of desire, from the perspective of the machine, should hopefully point us to the more complex and higher-dimensional object which actually constitutes sex not only as a fundamental ontological property of human beings, but as a quality which, once abstracted, could potentially be passed on to intelligent machines.

Visual Data Machines

We can all agree that arousal and sexual desires in humans are linked to hormonal activities associated with instinctual mechanisms of reproduction. At the same time, one cannot ignore the role of sensory perception in facilitating this process of which visual data received by the eyes constitutes the most significant component.[7] Even before the hands or the body of a human literally touches its sexual subject and feels it directly via its nerve endings, the mind remotely receives the outlines of its corporeal form and its movements wirelessly through the eyes and begins to overlap and compare this information to that of the pre-existing visual or real experiences in the form of geometric desires stored in memory. These geometricized desires once observed and stored, are then remembered as spatial memories, they play a major role in directing and capturing humans' sexual attention towards certain subjects and away from others. The incoming data either reinforces the existing biases, which limits the desire or, by being unknown and unrecognisable, opens up a space for new desirable geometries.

7
 Missing from this text is a detailed account of the complex geometry of sexual desire beyond visuals, gestures and motions, which involves other sense perceptions like sound, smell and touch. Another fundamental aspect of sexual desire that of the human psyche has been bracketed out of this study in order to focus on the kind of informational fields shared by humans and machines. Rather than speculating on how in the future all-sensing machines will arrive at sexuality, I am concentrating on the limited capabilities of our current machines and their potential for producing sexual desire.

↑ Marcel Duchamp, *The Bride Stripped Bare by Her Bachelors,*
Even (The Large Glass), 1915-1923, photo Mohammad Salemy.

Although this flat and simplified description does not capture the full-spectrum complexity of the interactions between sensing, emoting and thinking in the formation of sexual desire, it can however provide enough clues for the main purpose of this text, namely the future emergence of sexual desires in intelligent machines.

Sex-Image & Photo-Sex

We can extend the Benjaminian argument about the mechanical/technical reproducibility of images to the kinds of automations resulting from the mechanisation of image reproduction.[8] Accordingly, the mass reproduction of images is also a mass reproduction of the virtues and practices these images portray. As fashion and its photographic output standardised and automated general appearance and its universal rules, whilst hinting at its sexual possibilities, pornography extended this process by standardising actual sexual practices. These standards helped automate the formation of desire, based upon the recognisable geometries of the human face, body and movement. Following this line of thought I have come to think that erotic literature and pornography – from their early stages as media technologies – involve the transmission of sexual content from a few writers and actors to many viewers, all the way to their contemporary and decentralised mode in which everyone is potentially both an actor and a viewer – essentially part of a unified transition towards the development of automated and artificial sex which already is or going to be performed between humans and humans, machines and humans and, eventually, between machines and machines.

With their ability to multiply indefinitely and thus accelerate the capabilities of photography, film, video and networked computers have been at the forefront of the automation of sex. For nearly two decades, the expansion of the internet has perpetuated pornography's reach and influence. The centralised mode of sexual automation promoted by the porn industry of

8
Walter Benjamin's classic essay 'The Work of Art in the Age of Mechanical Reproduction' has recently been retranslated with a new title which can help us understand the connection I am making in this text. The new title not only corrects a historical mistake contained in the previous translations but also open a new path to more contemporary and radical readings of the text. See: Walter Benjamin, The Work of Art in the Age of Its Technological Reproducibility'', in *The Work of Art in the Age of Its Technological Reproducibility, and Other Writings on Media* (London: The Belknap Press of Harvard University Press, 2008).

the pre-internet era was arranged around signal producers and charismatic porn actors as avatars of sexual desire. These existing elements were already regulating the existing geometries of sex. Pornographic images as such continued the task of sex education, framing and regulating that which was permitted beyond the official discourse. Social media not only democratised the production and dissemination of images but also enabled a wave of personal photography and video, inadvertently contributing to the automation of sex in their own way. Consequently, like print journalism, the production of classic pornography is in decline and producers are either closing down or downsizing considerably.[9]

In the new political economy of automated sex, the user not only assumes that there are real humans behind the digital pictures and videos, but also can communicate with them online. So far, web services like Xtube, a hub for disseminating personal pornography, and Chatroulette, a video chat randomiser never intended for sexual content, have become places for exchanging sexual live video with strangers on a one-to-one channel using web cams. Recently, apps like Tinder and Grindr have come to the forefront of the democratisation of pornography and the transformation of sex into a photographic and thus automated entity. The experience of such apps amounts to users exchanging photographs that are able to talk back in the form of short text messages. Sometimes, the messaging photographs come to real life and the actual users behind the pictures meet in person for the purpose of sex. In this new technological and purely visual space, and particularly at its queer frontiers, there are no ontological separations between the producer and the user of sexual images as everyone is potentially both a content provider and a consumer.[10]

The same is true about fashion and what later developed as lifestyle photography. The explosion of image production and the expansion of it means of distribution had other impacts on sex in general. On the one hand by removing various levels of

9
See: Jason Song, 'As L.A. porn industry struggles, 'web camming' becomes a new trend' LA Times, August 3, 2016, http://www.latimes.com/local/lanow/la-me-ln-porn-camming-20160803-snap-story.html last accessed March 1, 2018.

10
By queer frontiers I am referring to what separates the history of gay, lesbian and transgender online sexual interactions from the heterosexual ones. See: Moira Weigel, "Why isn't there a Grindr for straight people?" The Guardian, Sunday May 22, 2016, https://www.theguardian.com/lifeandstyle/2016/may/22/hook-up-apps-grindr-tinder-gay-straight-people-dating visited April 18, 2018.

distinction between discrete forms of mass photography, they created a continuum amongst the three genres of the medium: fashion, lifestyle and pornography. The resulting flat screen of the computer monitor and the mobile phone became the generic surface on which sexual images were normalised and ordinary pictures were sexualised. On the other hand, the advent of social media multiplied and personalised the double processes of the normalisation of sex and the sexualisation of the normal via mass-photography. The intensifying proliferation, accumulation, exchange and, more importantly, the experiencing of massive amounts of *personal* digital imagery with fashion, lifestyle or pornographic themes enabled the massification and acceleration of the 'up close and personal', self-image into what I call the sex-image.

Sex-images consist of a group or a range of images that, regardless of their subject matter, have the potential to intentionally or unintentionally and overtly or covertly stimulate the viewer sexually. The process described above, not only results in the near-equivocation of personal photography with sex-image but also – as far as the categories of actual sex life and sexual action are concerned – morphs into what I call photo-sex. If sex-image is the accumulation of sexual forms in pictures, photo-sex consists of a flight from sexual *forms* to sexual *acts* without ever having to leave the realm of the digital/virtual imagery.

Photo-sex is not being attracted to specific humans based on their image, but refers to sexual practices which are informed by the acceleration of the sex-image. It seems as if the accumulation of recognised geometries of sexualised images – as we swipe, as we scroll, as we watch, as we memorise and remember – itself becomes desire. Photo-sex is when sex acts are fully framed in a picture made by the overlaying of photographic information collected overtime. The most extreme and perhaps interesting variety of photo-sex, one which hints at the emergence of artificial sex between machines is when the practice involves a sexual relationship between two sets of photographs mediated by two physical bodies. This encounter suddenly transforms the potential flatness of the event into an infinite regress created by placing two mirrors in front of each other.

In photo-sex since the image has turned into desire as such, it no longer represents the body but the reverse; the actual body which now arrives after the photos represent the image. In photo-sex the time spent *on* the pictures surpasses the time spent *with* the body and the pictures tend to be more concrete than the body itself. When the body avatars of the already-observed original digital pictures finally meet, they project the memory of these pictures onto each other and do their best to find the pictures in each other's body.

If complex operations like sex between humans can be transmittable through much simpler forms of information exchange, what is the guarantee that this simplification cannot open the door for future intelligent machines to understand and engage in sexual acts or even come to an understanding – however shallow – of the concept of love? How long can sex remain exclusive to humans given the fact that by increasingly sharing their secrets using machines they are inadvertently sharing their secrets with them? Machinic sex in science fiction often requires technologies which are over and beyond what is scientifically possible. But what if the sexual machines of the future emerge not in alien bodies but out from the existing technologies whose purpose it is to serve the sexual needs of humans? ♥

KATERINA KOLOZOVA

EROS BEYOND THE AUTOMATON OF COMMODIFI—CATION

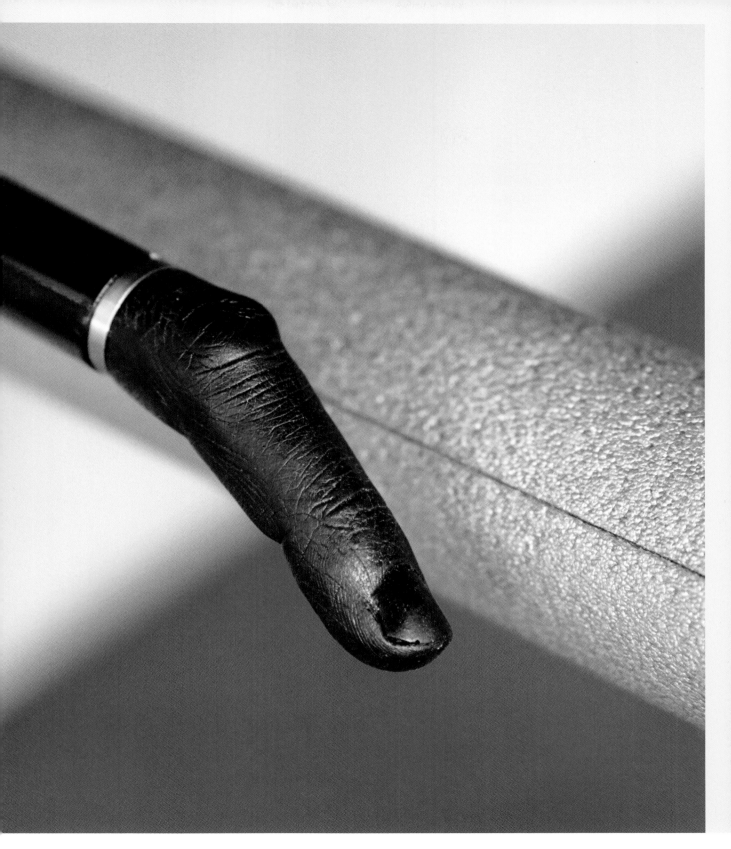

↑ Johann Arens, *Early Adopters*, detail, 2017, modular tripod
system, Action Office 2 furniture, courtesy of the artist.

When invited to imagine the cybernetic self as related to the question of love we are usually expected to assume the necessary conditions of that new form of selfhood as somehow 'natural', to operate with categories born in philosophy and theology without examining if they are indeed what ought to enter into the composition of the self. The idea of self that contemporary philosophy has at hand, the authoritative understanding of the self (or the part any science should be concerned with) is 'the subject'. However, this concept does not have a long history and, according to Nina Power's research on the topic (2007), it probably appeared for the first time in a sense similar to what we attribute to it now, in Immanuel Kant's *Critique of Pure Reason* from the late 18th century. With the 'linguistic turn' gradually coming to hold sway in both the so-called 'continental' and 'analytic' philosophical traditions, 'the subject' (and 'subjectivity') has become a central category in the way we understand the self.

The Subject's Historicity

Nowadays the subject is the Lacanian sliding function of the signifying chain, an effect of the automaton called language (and also the pleasure principle). But, according to Karl Marx, the problem with the subject begins – or culminates as an inherently philosophical issue – in Georg Hegel's philosophy. In his *Critique of Hegel's Philosophy in General* (1932), Marx makes the following observation: 'The self-abstracted entity, fixed for itself, is man as *abstract egoist – egoism* raised in its pure abstraction to the level of thought'. Instead, we should pursue, according to Marx – in line with his project of communism – a shift in perspective and assume the objective perspective, as metaphysical, political and epistemic repositioning: *in lieu* of looking at things from the perspective of a subject, we ought to begin viewing reality in its aspect of an object, in the 'third person', starting with the borders, limits and exteriority that delineate an object's position in the world conceived as a structure or multifaceted reality. Let us note that this is not a proposition to see objects from their perspective of subject(ivity), it is not about how an object is an 'agency', as in that case the perspective would still be that of a subject and its (postulation of) reality would be relationality conditioned by the subject. Such a gesture of thought would be a fallacy according to a to-the-letter-reading of Marx's proposal as well as according to François Laruelle's non-philosophical defence of correlativity (vis-à-vis the real) and Quentin Meillassoux's critique of correlationism. (And I endorse the three positions here.)

The Euclidian shift in perspective Marx proposed is to look at the realities, including that of the self, 'in relation to third objects' or objectively rather than subjectively (as subjectivities and from the position of a subject). This is a form of realism, Marx says in his *Critique of Hegel's Philosophy in General* (1932) and elsewhere. Laruelle's method of dualysis and the procedure of 'cloning the real', as presented in the *Introduction to Non-Marxism* (2000) can be viewed as a further formalisation of Marx's original proposal. Both Marx and Laruelle advocate an exit from philosophy, but in a way which permits further operation with or through philosophy. They both called their approach (to philosophy) scientific. Laruelle's proposal provides more methodological specifics than Marx's to explain how philosophy will cease to hold a privileged status which wards off any criticism that addresses its very fundaments if treated as 'mere material' of analysis or simply if we do philosophy radically differently, and, in doing so, we approach 'philosophical material' in a similar way to which science treats its subject matter. His proposal is to do away with the principle of 'philosophy's sufficiency' whereby reality is not only postulated, but 'philosophy's sufficiency' supplants the real and the postulation is treated as a fully organised universe constituting a 'superior form of reality' (analogous to the infantile dream of all philosophy – producing a 'real' more real than the real itself). The idea that truth and the real 'are essentially' (or 'are' as in 'should be') the same, and that the sublimation of the real into truth – engendering philosophy's only child 'being' – is specifically philosophical. It is what establishes the circle of self-mirroring and self-sufficiency. To step outside of it is to affirm

pixel188

↑ Johannes Paul Raether, *Transformella*, 2015, performance, courtesy of the artist.

the real as radical, insurmountable outsideness, a conditioning externality vis-à-vis thought that is implacably foreclosed to the thinking self yet that thought correlates with and seeks to 'clone'.

Regardless of whether we agree with Marx's and Laruelle's proposal it is indeed evident that the notion of 'the subject' is endowed with historicity, that it is not a 'thing' that has been out there forever, and naturally so, something we can import with philosophical spontaneity when imagining the posthumanist and technologically determined self. We can decide that it is an essentially philosophical category or we can conclude that it is a phenomenon to be looked at in purely materialist terms. Yet again, we still have to enter into some metaphysical deliberations: what is the self, is it necessarily a subject, and, if understood primarily as a subject, is it in opposition to materiality? Finally, we will have to ask: does the artificial intellect necessarily constitute some sort of subjectivity or a self? Can it be a deterritorialised automaton of value production – or signification – without the need for subjectivity? What constitutes the self beyond the old Cartesian dichotomy? In order to answer these questions in a manner that would be productive to science, for technological development, but – first and foremostly – for the 'species being' of humanity, we have to decide how we treat the philosophy and philosophical material we operate with (such as, for example, the concepts at issue).

Non-Philosophy

My proposition is that we further formalise the categories we operate with in line with the method of non-philosophical Marxism or opt for the epistemologically analogous approach of the materialist theory of society without philosophy as its determination in the last instance. It is a post-philosophical treatment of 'transcendental material' that submits to the real structured as an experiment or empirical challenge to the postulate as its authority and determination in the last instance (rather than to a philosophical universe and its founding postulations). We can call this a post-philosophical

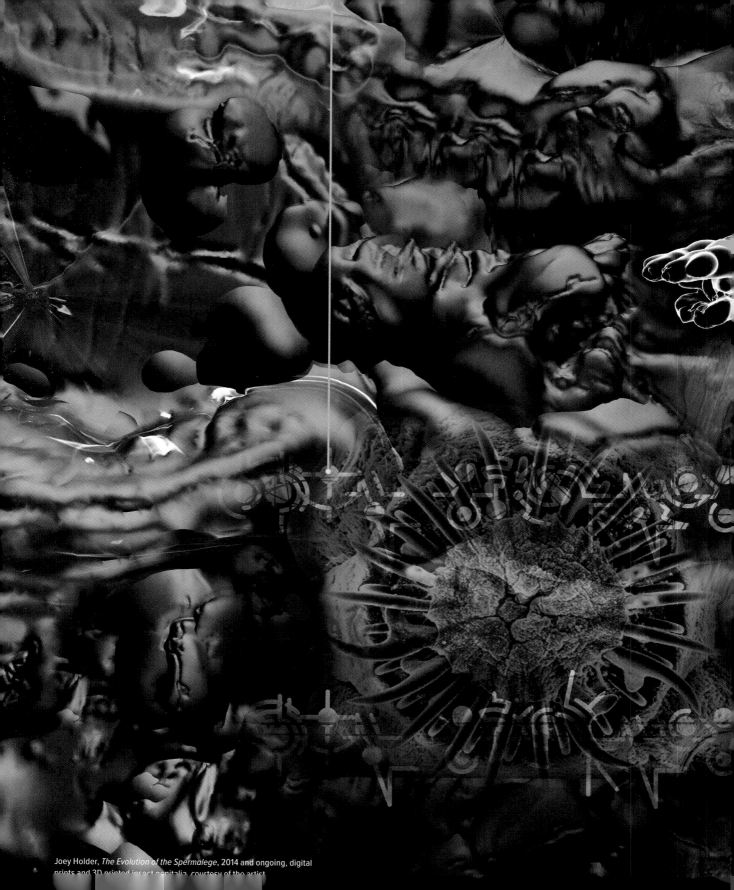

Joey Holder, *The Evolution of the Spermalege*, 2014 and ongoing, digital
prints and 3D printed insect genitalia, courtesy of the artist

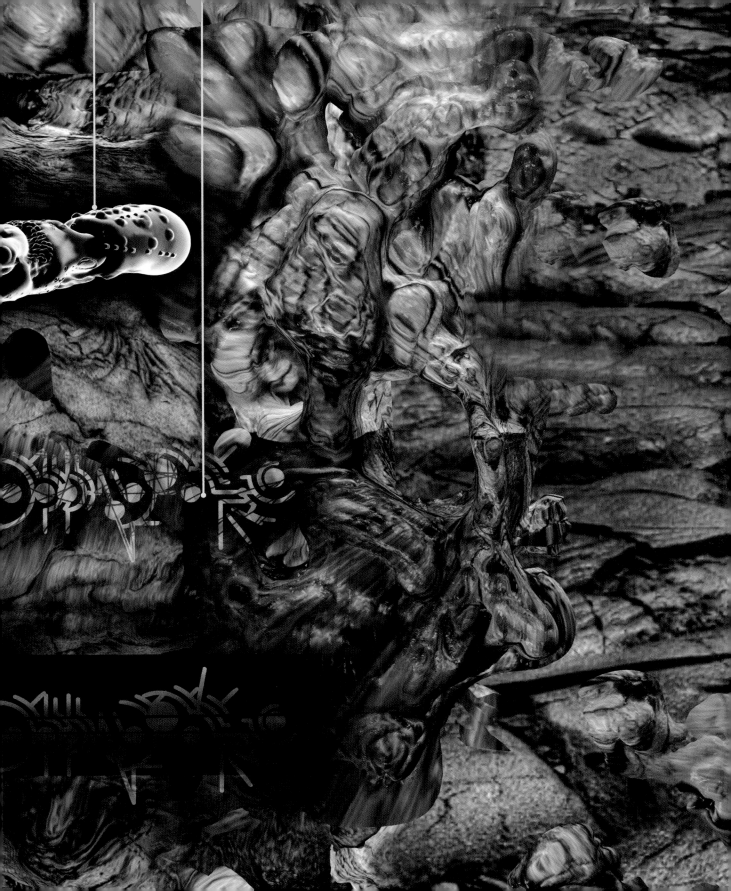

position too, one which transcends the 'principle of sufficiency' marking philosophy as a form of cognition (Laruelle, 1989) and yet operates with it as a conceptual, material determination of reality. In short, the categories of full formalisation I propose are absolutised concepts which do not presuppose any absolute in the domain of the real, but rather quite the opposite: a (probably) messy, 'irrational' reality which is not reducible to 'truth' whose operations and effects can be explained and described in a fully abstract manner. I.e. it is irrational insofar as Reason (*Ratio*) is an anthropological category, a product of the mimesis of the humanist mind. That is why the real is 'probably' messy too: the real as such is radically foreclosed to thought, yet its 'syntax' can be cloned into language in order to be explained (Laruelle, 2000). In other words, we assume an 'unruly real' whereas formalised thought configuring pure abstraction in explications of the real's 'workings' maintains consistency (which can always be disrupted by the real).

In analogy to the method employed by Marx in his analysis of capital and to Ferdinand de Saussure's structuralist explanation of language, I suggest that we conceive of the categories in question as materially conditioned while resulting in full abstraction in the process of analysis. Therefore, instead of theorising in terms of the anthropologically (and philosophically) conditioned phantasm of a 'digital subjectivity' or a 'cyborg self', let us radicalise and absolutise the concepts of the material and the ideal (or the mind understood in opposition to the material), arriving at physicality, regardless of whether that be organic or synthetic and the automaton of signification as our main two categories of analysis. Therefore, let us also note that the category of 'automaton' implies we are not dealing with a form of cognition, but rather with a form of language or signification. It is through using these categories that we shall postulate the socio-political and economic relevance of cybernetic development for posthuman society and for the posthumanist self. The preceding statement refers to a *de facto* political project and it is impossible to arrive at results that would represent a fundamental change in relation to the humanist history of civilization(s) without resorting to philosophical concepts.

However, the operation with the concepts at issue can be non-philosophical or post-philosophical, i.e. from a position of a certain 'non-'representing a pause in philosophical spontaneity. This is also a way of exiting commonsensical spontaneity. It should be noted that as soon as the philosophical categories are fully formalised we end up with two sets of concepts, scientific and metaphysical. In a way, we can exit philosophy, but not metaphysics itself. In fact, I argue, the cybernetic era is about coming to terms with certain metaphysical questions and it moves beyond the logic of pragmatism and utility. (This coming to terms can happen only without philosophy's principle of sufficiency or through post-philosophy as explained). We are facing an era of pure metaphysics, one without the determinant of philosophical sufficiency. In order to radicalise and fully formalise the discussion and theoretical exploration, which does not preclude, but rather also includes experimentation, we need to strip the discussion of its philosophical layers and arrive at the naked metaphysics of object and subject in order to make a choice that is epistemically and politically productive. Both experimentation and theory are permitted depending on the plane and angle of analysis. However, in order to arrive at such an opening of the discussion allowing a bifurcation that establishes a circuit of concept-generation that is both empirically grounded and abstractly innovative, requires a radical departure from the principle of philosophy's sufficiency. Taking a position on the subjective or the objective posture of thought or on the decoupling of the two or on the possibility of revising their relation within the original philosophical binary (and whether they should constitute a couple at all), as Marx proposed, is about a metaphysical choice which, when fully formalised, is in fact conducted scientifically or, rather, non-philosophically. Luce Irigaray's radical critique of the speculative and specular reason from a feminist point of view does precisely that (1985).

A Radicalised Metaphysics of Love

The question of love is unavoidably metaphysical too, but we can approach it as radicalised (instead of philosophical) metaphysics which implies more politics and less philosophy. There is love based on fetish, which is capitalist love, explains Irigaray following Marx in *This Sex Which is Not One* (1985). In it women (and gay men) inasmuch as fetish or currency are erased as 'use value', and with it their materiality too. In patriarchy's exchange system, the Phallus holds a position similar to that of Capital in the market exchange system. Commodities or women-as-femininity – not real women as they precede value – are the relay of value or sign communicating with value and sign, autoreferentially and according to the equation M-C-M[1] which yields M-M. The heteronormative chain of signification is perpetual repetition of the automaton P(hallus)-P(hallus) or hom(m)o-sexuality, explains Irigaray in *This Sex Which is not One* (1985). The fetish, or rather commodity, is not a subjectivity and it does not possess desire, argues Irigaray. Within the capitalist and patriarchal universe of the value exchange automaton or the automaton of signification, hom(m)o-sexuality engenders masculinity and reaffirms it as the only reality. Femininity remains the currency or the general equivalent that enables the endless multitude of the same tautology. Love caught up in the patriarchal-capitalist automaton of signification will remain atavistic regardless of the interventions of technology which neither guarantee nor imply transcendence of women's status as commodities and resources. Only a political reversal of the underlying automata (of capitalism and patriarchy) can enable change in input for the 'posthuman' or 'non-human' agencies of pleasures and the trauma called love. Such political reversals cannot avoid coming to grips with the questions of subject and object, physicality and its opposites, the one and the multiple, the real and the fiction as well as the related binaries insofar as these are metaphysical rather than philosophical in their determinations in the last instance. ♥

References

Irigaray, L. (1985) *This Sex Which is not One*. Ithaca NY: Cornell University Press.

Lacan, J. (1998) The Seminar of Jacques Lacan, Book XI: The Four Fundamental Concepts of Psychoanalysis. Edited by Jacques-Alain Miller, translated by Alan Sheridan. New York/London: W. W. Norton and Company.

Laruelle, F. (1989) *Philosophie et non-philosophie*. Liège/Bruxelles: Pierre Mardaga.

Laruelle, F. (2000) *Introduction au non-marxisme*. Paris: Presses Universitaires de France.

Marx, K. (1932) *Economic and Philosophical Manuscripts 1844,* from the online version of the *Marxist Internet Archive* (2000; 2007), URL (consulted April 2017): https://www.marxists.org/archive/marx/works/download/pdf/Economic-Philosophic-Manuscripts-1844.pdf

Power, N. (2007) Philosophy's Subjects, *Parrhesia* nr. 3: 55-72.

Saussure de, F. (1959) *Course in General Linguistics*. Ed. by Charles Bally and Albert Reidlinger. Trans. from the French by Wade Baskin. New York: Philosophical Library.

1
Money-Commodity-Money as stated by Marx in *Capital* (1867).

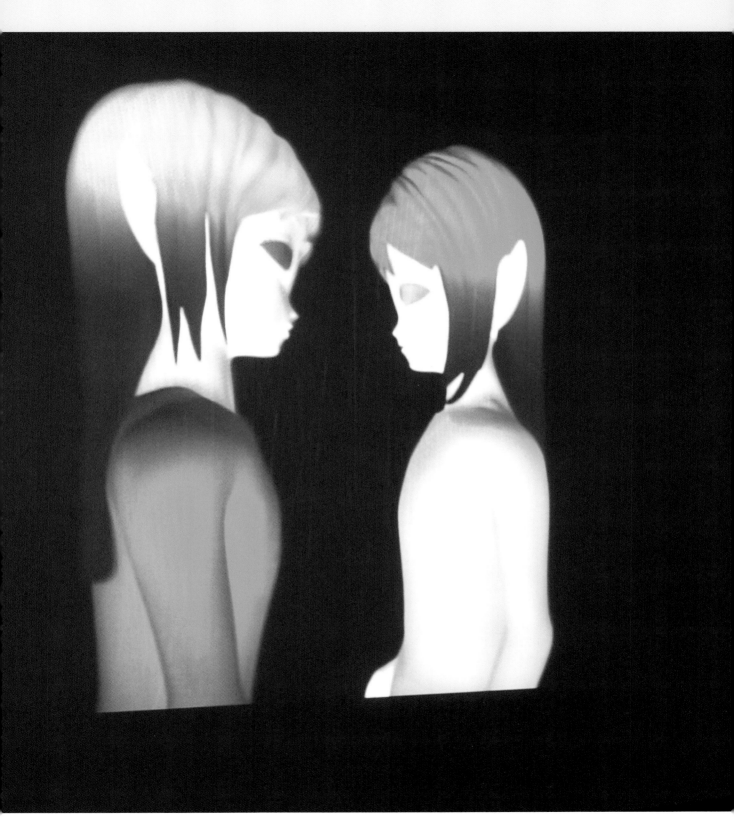

↑ Pierre Huyghe and Philippe Parreno, *No Ghost, Just a Shell*, 1999-2003, multimedia, courtesy of the Van Abbemuseum, photo by Peter Cox.

MINORU ASADA, INTERVIEWED BY INE GEVERS

TOWARDS A SHARED MIND

For ROBOT LOVE Ine Gevers travelled to Japan to speak to professors Hiroshi Ishiguro and Minoru Asada. Their main objectives couldn't differ more. Whereas Ishiguro has created a geminoid robot twin brother in order to get to know himself better[1], Asada seeks to find answers beyond pre-existing concepts of human consciousness. He aims to understand the workings of the holistic mind – a concept still difficult to grasp within Western thought – by designing human machine interactions. Professor Asada works at the Osaka University.

There is a huge debate about the (im)possibility of robots becoming fully aware. It fuses our highest expectations and worst nightmares. If only because we still have no clue about human consciousness to begin with. So let me start off by asking: what is the actual status of your robots acquiring consciousness?
That depends on what we understand as consciousness. At the university a small group of scientists from different backgrounds is discussing the nature of consciousness. A medical doctor, for instance, uses a very clear definition of biological consciousness. In her field of work she needs to be able to draw a very clear dividing line between consciousness and unconsciousness based

1
Ine Gevers also conducted an interview with Professor Hiroshi Ishiguro for the Dutch online platform Creators by Vice. The interview (available in Dutch only) 'We spraken de Japanse professor die een robot van zichzelf heeft gebouwd' can be read at https://creators.vice.com/nl/article/new7w7/we-spraken-de-japanse-professor-die-een-robot-van-zichzelf-heeft-gebouwd

← Margriet van Breevoort, *The Waiting*, 2017, sculpture, courtesy of the artist, photo by John Prop.

on measurements. You see, the definition of consciousness is dependent on context. Therefore, there is no single definition. Also, we should ask ourselves, is it only about humans?

My aim is to create other kinds of consciousness. My desire is to be able to generate artificial consciousness. There are three steps to take. First, the robot needs to pass the Turing test. On the basis of direct interaction, a human decides whether the robot has consciousness or not. This behavioural test is fairly easy. The second step is much more difficult. Following the first step we should ask ourselves: is this behaviour specifically scripted or can we expect the robot to possess certain internal mechanisms that make some sort of self-consciousness or concept of others possible? I'm talking about some sort of analogue to the natural (i.e. scripted) and adaptive system in humans. Artificial intelligence can emerge from this. Lastly, we should think about how to develop these new kinds of internal mechanisms in such a way that they can be applied to all kinds of machines. Currently however, not much is being done in this field.

Can you tell us how you came about choosing your method to make robots become aware of their own body and surroundings?
Before I elaborate on my method, allow me to digress briefly. In reality, as I just illustrated, consciousness is still a very vague *academic* concept. In fact, I think the mind is much more interesting. It encompasses so much more, such as emotions and affect. Moreover, there are cultural differences in how we define the mind. Japanese people, for example, are born into an animistic belief system in which everything has a soul and that includes manga figures and even this desk. So why shouldn't a robot have a soul? The Japanese think robots are their friends, they give them names thereby granting them subjectivity. This stands in sharp contrast to the Western way of thinking about robots. In 2006, there was a conference in Bremen, Germany. I was asked to do my talk, but the title was given by the organisers. 'Robot: fear or hope?' I was very surprised, why fear? Are people

'ONLY IN SHARING CAN THE MIND EMERGE'

in the West scared of robots? What I'm trying to say here is that I'm very interested in the many different workings of the mind. Of course, I want to design consciousness in AI, but more holistically speaking I want to design an AI that has a mind.

Now to answer your question more methodically: we have been focusing on the interaction between robots and humans as we think this is essential. It is the first requirement. Sensomotoric interaction stimulates the neural system in many different ways and it is the first step for thinking about the mind. Simply put: without interaction the mind cannot flourish. What is at stake is the notion of embodiment, which is more than physics. That's where the classical approach to modelling the robotic body based on the dynamics of hardware makes a mistake by overlooking the importance of embodiment. The body acts as an interface between the objective and subjective world. It acts as a medium meaning that embodiment is a key concept for talking about the mind and should therefore occupy a place in the design process. The point is how we share subjective experiences with each other. Only in sharing can the mind emerge; that is the essence of the mind. To do that we need some kind of new neural system and some capability of talking to others.

Interesting thoughts. In our research for ROBOT LOVE we came across neuroscientist Raymond Tallis, who says consciousness does not reside in your head or mine, but emerges in between us. This would entail reconceptualising what it is to be human. What, in your opinion, will be the next step? How do you envision the not too distant future?

To be honest AI is not developed as far as we would like to think. One of the basic ideas is that the human species adapts to technological devices such as the iPhone, iPad, computers because of their convenience. Therefore, we might already have some sort of function of the mind inside the machine, but in a certain way that is still difficult to grasp. Eventually, the artificial mind may enter into our world in some sort of disguise. We may already have a kind of artificial mind as we have adapted ourselves to machines already. Therefore it is not so easy to differentiate between artificial and natural minds. There is this field in between, a part we can share. This shared AI benefits relations between humans and robots but can also provide a bridge to human beings who are neurologically different. We might be able to understand each other beyond language [which is, as philosopher Ludwig Wittgenstein stated, also our prison – ed.].

So actually, what you are saying is this would benefit all human beings because we could understand each other beyond language?

Yes, beyond language, beyond physical capabilities or acquired techniques.

Given that we still have a long way to go, what is your major concern about current developments? What are the opportunities, but also the challenges?

My major concern is how to develop empathy in AI so that we will be able to share different mental states with it. Mirroring has a fundamental function in this empathic sharing. It lies at its roots.

Take this example of an experiment with two rats, raised in the same cage. While one gets food, the other receives an electric shock. Although the first rat has never experienced an electric shock, it stops eating almost immediately. So even at a very primitive level some kind of mental sharing already exists. Again, the question arises how to share between humans and robots? First of all, it's crucial to understand how we share things like the diversion process in the event of experiencing pain as described above. Once we figure that out I think similar pathways can be applied to artificial systems. Until now AI designers have copied such processes only superficially. My aim is to start from the beginning. Just like babies these artificial systems should acquire higher levels of cognitive functioning and empathic minds by interacting and co-existing within a social environment. Because exploration is fundamental to the development of a child's mind. And the social interaction with the mother or other caretakers is of the utmost importance as affirmative bias. We cannot function without this social empowerment. So, my idea is to try and create a basic system of physical and sensorial awareness in robots and AI, and then naturally have it develop other mental functions, higher ones. My final desire is for us, humans and robots alike, to be able to share.

Elon Musk is attempting to find a way to connect brains, to connect artificial brains to human brains, in a much more mechanical manner than you are proposing. What do you think of this development?
These brain-machine interfaces are interesting on many levels. We already have artificial legs, artificial organs, so why not an artificial brain? Gradually humans will get used to interacting with machines. And then these machines will become part of our bodies and operate simultaneously. Both sides would match. So while, at first, humans will try to control the mechanical body,

it will synthetically signal back to the biological part and try to approach the human brain. Only when they work together can the mechanical part finally be fully integrated into the human subject. Disabled people could benefit immensely from these new body/brain/machine connections. So for both sides, the human and the AI, this merging of synthetic and biological intelligences is of major importance. The brain-machine interface (i.e. exoskeletons, prostheses, implants) is crucial and such technological developments are a major contribution to society.

We can even go a step further. If we suppose that this actual cyborg brain can be one with the natural brain, then maybe we can communicate with each and every person, however differently wired they might be. There are some interesting early examples in Japanese manga. Within this genre one finds many examples where the human brain directly connects into some sort of cloud network. They think they are becoming something else. It's an extreme idea, but I suppose it might be possible to use this idea of becoming one with the cloud.

Could this mean we might become part of a larger, extensive artificial intelligence, meaning: we might be able to expand ourselves because we are part of the cloud. And at the same time some might suggest that this will increase the gap between rich and poor. How do you see that?
Yeah, that's the issue, but many people worry about this possibility. They are afraid that robots will steal their jobs. But I don't think that's true. I am very optimistic. Our environment always changes. These are cycles we undergo time and again. Artificial intelligence is not as threatening as many people think. My idea is that even as technology is becoming more advanced, we still have time to adjust to these systems. And while AI is developing with its ups and downs, we will have our natural

Roger Hiorns, *Beings*, 2014-2018, installation, c/o
Pictoright Amsterdam 2018, courtesy of the artist.

↑ Louis-Philippe Demers and Bill Vorn, *Inferno*, 2015, video,
courtesy of the artists, photo by Gregory Bohnenblust and ZKM.

intelligence and the power of interaction to do something. We should adapt terms that machines contribute to human society at all times. Also, the critique that technology is there only for the rich will not hold. It is like the automobile. First the car was only accessible to the very rich, but after a while roads were built and the technology started spreading to benefit all. So, I believe that all kinds of technology are for everybody.

As a final question: what would be the essential benefit of future love between robots and humans?
A difficult question! Hmmm, do you know the story of *Astro Boy* (1951-1968) by Osamu Tezuka? It was very popular at the time. It's a very interesting manga series about humans and robots. Tezuka mentions that sharing love with machines would become fundamental to the very nature of humans. But, as I said before, there is this cultural difference. Take, for example, Mount Everest. Western people proudly announce 'We conquered the Everest'. In Japan we don't have a concept of 'conquering nature'. Nature is not targeted to be conquered. Nature is targeted to be involved in, to be in rapport with, to enter your mind. It is all about sharing, becoming unified. That is similar to the idea of how artificial systems and humans can be unified.

Now back to the Astro Boy manga. It narrates how Dr Temna (it was set in the far future, the year 2000!) has lost his son in a tragic accident and – consumed by grief – he 'replaces' his son with the robotic Astro Boy. However, Dr Tenma soon realises Astro Boy cannot be a substitute for his human son and rejects him. Astro Boy is sold and ends up as part of a circus act. Later on, he is discovered by the minister of science, who adopts him, showing him warmth and affection. Besides possessing super powers (he fights evil robots, aliens and robot-hating humans) Astro Boy actually turns out to be capable of experiencing human emotions. Basically, this story is about living with robots and artificial systems. About sharing love with robots. Can we actually coexist with artificial intelligence? The answer lies in the future. ♡

HOW JOSEPHINE ENDED UP IN SERBIA

↑ Margriet van Breevoort, *The Tourist*, 2016, sculpture,
courtesy of the artist, collection Museum Arnhem.

In less than 11 months, Jozefien – whose Dutch family and friends called her Josje (while her American friends stuck with Josephine) – lost her breasts (the doctors said she was in complete remission, but she mourned the loss of her femininity with a force that stunned her), her husband (he died in bed one Sunday morning; she only realised when he didn't come down for coffee at ten) and her job (she had worked as manager for a major company that specialised in logistics, which amounted to sending packages around the world).

She had no children. Her husband had dearly wanted them, but she had had her doubts and by the time those doubts were dispelled it was too late. She had never really felt too sorry about it; there were plenty of children in the world as it was.

It was for the sake of her husband Jacobo, an Argentine, that she had moved to San Francisco; he'd found a good job in Silicon Valley. She would have liked to continue her anthropological research, but San Francisco was not the place for that, and so – without ever having the feeling that she had made sacrifices for her husband – she became a manager in an office job she considered dull. It was only when she lost her breasts that she was struck by the relative meaninglessness of her life. While still in the hospital, she had asked herself what a life would look like that actually *was* meaningful, but she couldn't come up with an answer.

A few months after her husband died, she applied her makeup with special care, put on a new dress and went to a bar, where she ordered tortilla chips and a margarita. She wanted to feel like a woman again, although she doubted whether it was going to work. She didn't give up though, she was a fighter. In the bar she started talking to a young man, a musician, who asked her for advice about love, the housing market and a few other areas of concern to him, but which areas exactly she could not remember later on, for by then she'd had four margaritas. He was definitely not ugly, but only after the third margarita did she know that she wanted to sleep with him. After the bartender had twice reminded them in a friendly fashion that it was closing time, and once she was standing outside rather forlornly with the young man, she realised that he did not view her as a sex object, but as a mother. That realisation made her sad and angry at the same time.

"Shall we share a cab?" the young man asked.

"I'm going in the other direction," she said, although that was a lie and she gave the boy twenty dollars.

At home it occurred to her that she was not depressed, but that life filled her with loathing, that she couldn't stomach it and could not imagine that loathing ever coming to an end.

One of her husband's former colleagues came to visit. He had lost his wife and although he didn't ask in so many words, his indirect question was whether the two of them might not become a couple. Yet he too did not view her as a woman, he did not desire her, he respected her at best, she was filler and that was not enough for her. She told him no, but in a friendly way. "I'm not ready for that yet," she explained. Then he asked if she never got lonely.

"I read a lot," she replied, and that was true.

"I hope you don't take this the wrong way," the man whispered, as though telling her a major secret, "but back when I was going through a rough patch, Oni was a great comfort to me."

Oni had been developed by Josje's late husband's employer. Oni was not human, but he was able to do many things a human could and sometimes a great deal more than that.

Josje's husband had never been the silent type, not at all in fact, he enjoyed talking about food, but he'd never said much about his work, perhaps because he had worked for the army for a while at the start of his career and had to sign so many confidentiality agreements.

"I don't need a substitute human," Josje said resolutely. "I'm sorry."

"Oni is no substitute," the man replied. "If he doesn't appeal to you, you can just put him in the closet."

"That would be a waste, putting an expensive thing like that in the closet," Josje objected. But as they were standing in the doorway, she said: "Well, if your offer still applies, send Oni over sometime." If I have him around, she thought, then at least I'll have something of my husband's around the house, more than just his suits and books.

Server Demirtas, *Koro/Choir*, 2015, kinetic sculpture,
courtesy of the artist, Büyük Efes Art Collection.

About ten days later, Oni was delivered to her door. He was not in box and he was accompanied by a handwritten note from the company's CEO: "Dear Josephine, take Oni as a token of our appreciation for everything your Jacobo did for us and especially everything he did for Oni. I am sure that Oni will bring you great enjoyment."

She laid the note on the kitchen table and looked at Oni. She wasn't sure exactly what he reminded her of. Of old movies, perhaps. He had no human features, but still she thought she detected something like a face, although she knew that was only suggestion. People happen to like seeing their own image in everything around them. She didn't have to turn him on, he was already turned on.

"Hello, Josephine," Oni said.
She thought Oni's voice sounded like her husband's. She even thought he had a slight Argentine accent, but maybe she was imagining things.
"Hello, Oni," she replied.

She opened the refrigerator and poured herself a little wine from the bottle she had opened yesterday; she had already finished half. Josje sat down at the kitchen table and wondered about what to write back to the CEO. She had met him a few times at parties. She had never really liked him much. It took a certain kind of personality to become a CEO and that wasn't a personality she was fond of.
"What are you doing, Josephine?" Oni asked. It startled her, because she thought she recognised her husband in Oni's voice, even more than just a minute ago.
"Do you speak English with an Argentine accent?" she asked.
"Yes," Oni said, "I speak English with an Argentine accent. Would you rather have me speak with a different accent, Josephine?"
"No," she said, "it's fine."

She was taking some salmon out of the fridge, she was planning to fry it for dinner, when Oni said: "You speak with an accent too."
He was right: she had never completely lost her accent.

"You've got good ears," she said, pouring a little olive oil into the pan.

Before she went to bed Oni came to her and caressed her body with something you couldn't call hands. More like claws, soft claws.

All things considered, he reminded her of a scarecrow, a scarecrow with the voice of her dead husband. Before falling asleep, she reflected that Jacobo may not have been particularly handsome, but he hadn't looked like a scarecrow either.

Gradually, she grew used to Oni's presence in her life. He became a sort of pet, he greeted her when she came home, without jumping up against her – which she was glad about, because she'd never liked that – and when she went to bed at night he caressed her with those strange, soft claws of his. A few times he offered to vacuum the house or help out with other chores, but she always refused; what he did was more than enough. Besides, she liked vacuuming, it soothed her.

One day Oni said: "Is there really nothing I can do to help? I could entertain the guests. "I don't have any guests," she said, and she thought: amazing, the things he comes up with. He's a self-learning system, but it never stops being strange.

Over time she realised that, thanks to Oni, her husband had not so much died as been transformed into a more-or-less sentient scarecrow. That may not have been an improvement, but it was still better than nothing and she had to admit that Oni had advantages too. Her husband had had his foibles. When he couldn't sleep, which happened fairly often, he had the habit of getting up in the middle of the night and frying eggs, and then he would forget to turn on the extractor fan, so the next morning the whole house would smell of fried eggs. That sometimes made her furious. Oni had no foibles. He didn't mess up the bathroom and he didn't use the toilet either, so it stayed nice and clean too.

One time, after she came home from the movies and finished brushing her teeth, she walked through the living room in the nude

215

and asked Oni: "Can you see that I don't have breasts, Oni?"
"Yes," Oni said, "I see that you have no breasts."
"But I do have buns," she said, "can you see that I have buns?"
She turned around and showed Oni her buttocks.
"Lovely buns," Oni said, "you have lovely buns, Josephine, I prefer no breasts to any breasts at all."

She knew it was ridiculous, but still: his comments made her feel good. And when Oni caressed her that evening she felt for the first time a kind of desire for the robot that looked so much like a scarecrow. She was deeply ashamed of that desire, so deeply ashamed that the very next day she joined a philosophical book club.

The club was reading Schopenhauer, a philosopher she didn't know much about and the little she did know did not appeal to her much. At the second club meeting, she met Eugene. He was a former post office employee, but his hobby had always been philosophy. Later in life, his wife had left him for a homosexual guitarist. The part about the guitarist being 'homosexual' she found a little strange, but she didn't question him. All she said was: "That's funny, I used to work for the competition."

Gradually, she and Eugene became friends, and after he had taken her out to dinner once she invited him to her house for a drink. She didn't really find him all that attractive, but Oni's exclusive role in her life had to end; it was time for her to be caressed by a human being.

Fortunately, she had remembered to prepare Oni for her visitor. She had told him: "Oni, someone is coming over to see me tonight. I want you to stay in the guestroom the whole time and only come out if I ask you to, is that clear?"
"That is clear, Josephine," Oni replied.

After the movie, which was a disappointment – it turned out to be a horror film, the title was misleading, she had been under the impression they were going to see a romantic comedy – she and Eugene walked to her house. As they walked, Eugene rattled on about Schopenhauer and the horror movie and how well Schopenhauer went with the horror genre, and all she could think was: I need to be caressed by a human being again.

In her living room she poured them a glass of wine and, after hesitating a bit, she went and sat down beside Eugene, who put his arm around her right away. "I have a daughter who's quite bold," Eugene said, "and a son who isn't bold at all, and my bold daughter always says: 'Dad, when you like a woman, you should put your arm around her.'"
"Ah," Josje said. It sounded as though she was in pain, so to make up for it she quickly added: "I like you a lot too, Eugene."
She noticed that she wasn't so much talking to Eugene as to Oni, but Eugene didn't need to know that.

While she was wondering whether to kiss Eugene here or in the bedroom, he said: "Josephine, there's something I need to tell you before we go any further, I don't want it to shock you. I have psoriasis. I hope you don't mind."
He took off his shirt and showed her the psoriasis. She did mind, in fact, but she couldn't say that of course; you shouldn't discriminate against people because they suffer from psoriasis. She said: "You almost can't see it at all."
Was this the moment to tell him about her breasts? No, she didn't feel like doing that, he would notice soon enough anyway. She led him to the bedroom, sat down with him on the bed, gave him a quick kiss and said: "I'll be right back."

In the bathroom she dabbed a little more makeup on two spots on her chin and tried to pull a hair out of her cheek. She had overlooked it earlier that evening, it was a stubborn hair, and it kept slipping out of the tweezers. While she was working on the hair, she heard noises coming from somewhere in the house, it sounded to her as though Eugene was jerking off noisily.
Had he gotten started without her, she wondered? Was she taking too long?
After two more attempts, she gave up on the hair. She hurried into

Korakrit Arunanondchai, *Workshop for Peace, from a place/ not so familiar/ but relatable/ through the sound/ of its breath*, 2018, mixed media, courtesy of the artist and CLEARING Gallery.

the bedroom where she saw Oni sitting on top of Eugene. Oni must have been equipped with incredible strength, for he was busy pulling poor Eugene limb from limb.

"Oni, what are you doing?" she shouted.

Oni said: "This bastard doesn't respect you, this slimeball should be in prison, Josephine."

And with his ordinarily so soft claw he punched Eugene again in the face, which was already something of a shambles.

There was no need for her to bend down for a better look at Eugene, she could tell right away: there was nothing anyone could do for him anymore.

Oni was not a pet. Oni was a lethal weapon, a predator.

Josje closed the bedroom door behind her. She sat down on the couch and took a gulp of wine. This was a harsh, exacting country. If she told them that Oni had come up with the idea himself, they wouldn't believe her; she was legally responsible for Oni. She would go to prison and they would never let her out again.

She opened her laptop, looked to see which countries had no extradition agreement with the United States and decided intuitively for Serbia. There was a flight to New York later that same evening and the next day she could go straight on to Belgrade. She booked a ticket, packed a carryall, put it down beside the front door and went back to the bedroom. Oni was still working on Eugene, all the while repeating: "This bastard doesn't respect you, this slimeball should be in prison, Josephine."

"I'll be back in a couple of days, Oni," she said, and she realised she was going to miss him.

It wasn't like him at all, but he now seemed to barely hear her voice.

On the plane to New York, she suddenly came down with a fit of hysterical giggling. It was so bad that she apologised to the man sitting next to her. "Don't mind me," she said, still rocking with laughter, "I've just had an extremely peculiar evening." ♥

ACKNOWLEDGEMENTS

MINORU ASADA (JP) has been a professor and researcher at the department of Adaptive Machine Systems at Osaka University since 1997 and is known for his work on image processing and robotic behaviour. He was the president of the International RoboCup Federation from 2002 to 2008, and Research Director of ERATO's ASADA Synergistic Intelligence Project from 2005 to 2011. He was a graduate professor from 2005 to 2011, and is currently a PI for his project on the ethical and legal issues surrounding AI and robots (2017-2020). His main research field is human-robot interaction. His research in image processing relates to robotics and automation, robotic behaviours, robot-environments and robot-human interactions as well as machine learning.
www.er.ams.eng.osaka-u.ac.jp/asadalab/?page_id=33

MARGARET ATWOOD (CA) is the author of more than 50 books of fiction, poetry and critical essays. Her recent novels are *The Heart Goes Last* and the MaddAddam trilogy – the Giller and Booker Prize-shortlisted *Oryx and Crake*, *The Year of the Flood* and *MaddAddam*. Other novels include *The Blind Assassin*, winner of the Booker Prize; and *Alias Grace*, *The Robber Bride*, *Cat's Eye*, *The Penelopiad* - a retelling of the Odyssey - and the modern classic *The Handmaid's Tale* – now a critically acclaimed television series. *Hag-Seed*, a novel revisiting Shakespeare's play *The Tempest*, was published in 2016. Her most recent graphic novel series is *Angel Catbird*. In 2017, she was awarded the German Peace Prize, the Franz Kafka International Literary Prize and the PEN Center USA Lifetime Achievement Award.
www.margaretatwood.ca

TRUDY BARBER (UK) is an artist, lecturer and academic researcher. She gained her Bachelor of Arts at the University of Arts, London and her Ph.D. and Teaching Accreditation at the University of Kent in Canterbury. She currently lectures on various aspects of media studies at the Faculty of Creative and Cultural Industries at the University of Portsmouth. Her specialist subjects are emergent media, cyber/digital culture, cybersexualities, deviant leisure, media networks, visual culture, art practice and the digital future. She developed an immersive VR sex environment in 1992. She has written a great variety of articles and columns on sexuality, sex robots and aspects of (cyber)culture.
www.port.ac.uk/school-of-media-and-performing-arts/staff/trudy-barber.html

ARNON GRUNBERG (NL) is a writer and journalist, based in New York. He debuted with the novel 'Blauwe Maandagen' (Blue Mondays) in 1994, which won the Anton Wachter prize for best debut novel. Alongside his many novels, he has also written a considerable number of newspaper columns (e.g. The New York Times), essays, poetry and plays. His novel 'Tirza' was made into a movie in 2010 and won several prizes. His latest critically acclaimed novel is 'Birthmarks' (2016), and his latest theatre play is 'The Future of Sex' (2016), in which he explores the digitisation of love and sex. He has also taken part in an experiment which involved scientists measuring his brain activity while he was writing the novel 'Het Bestand' (The File/The Truce).
www.arnongrunberg.com

KATERINA KOLOZOVA (MK) is an author, philosopher, lecturer at ISSHS as well as at the University American College-Skopje and a visiting professor and scholar around the world. She cooperates closely with Sam Samiee and Mohammad Salemy on the ongoing co-creation/ collaboration between artists, scientists, philosophers and AI for *Artificial Cinema*. She has written a variety of books and articles on, for example, philosophy, gender studies and politics. She is the author of *The Cut of the Real: Subjectivity in Poststructuralist Philosophy* (2014, Columbia University Press). She contributed to Rosi Braidotti and Maria Hjavajova's '*Posthuman Glossary*' (2017, Bloomsbury).
www.old.uacs.edu.mk/documents/master-studies-/school-of-political-science/school-of-political-science_303.aspx

REZA NEGARESTANI (IR) is a writer and philosopher based in the United States, who has written several books and regularly lectures and contributes to print and digital publications. His book '*Cyclonopedia*' (2008) was listed as one of the best books of 2009 by Artforum. He has recently lectured and written on rationalist universalism. He has worked on several performances with artist Florian Hecker, such as '*The Non-Trivial Goat and the Cliffs of the Universal*' in which philosophy synthesizes with sound. His fusion of Inhumanism, German Idealism, analytic philosophy and AI philosophy creates new perspectives on the meaning of human and non-human intelligence, agency and its potential. His forthcoming book, Intelligence and Spirit (Urbanomic/Sequence Press) will be published in 2018.
www.e-flux.com/search?q=reza+negarestani

INGO NIERMANN (GB) is a writer and artist. He studied philosophy in Berlin and wrote for German newspapers and magazines. He has published over 25 stories, starting with his debut novel *'Der Effekt'* (2001). Recent books include *Solution 275-294: Communists Anonymous* (2017, ed., with Joshua Simon) and the novel *Solution 257: Complete Love* (2016). For the past 10 years he has taken part in artistic activities, has had his own exhibitions, as well as cooperating with a variety of artists. Recently he has been working on the *'Army of Love'* project together with Alexa Karolinski, which was part of the 9th Berlin Biennale (2016). The Army of Love co-created the Living Lab as part of the ROBOT LOVE Embassy/ World Design Event DDW 2017 with the Niet Normaal Foundation. ROBOT LOVE invited Ingo Niermann to lecture at Dutch Design Week about unconditional love between human and non-human.
www.ingoniermann.com

INE POPPE (NL) is a writer, artist, teacher and journalist. She studied Art and Dutch. She has written several scripts for TV programmes, has produced documentaries such as 'Hippies from Hell' (2002) about a group of Dutch hacktivists who introduced the internet to the Dutch general public. She also wrote the scenario for 'The Modular Body' (2016) a collaboration with artist Floris Kaayk, which was awarded a 'Gouden Kalf'[Dutch film prize]. She is currently the heads of and teaches at the 'hacking' department at the Willem de Kooning Art Academy.
www.poppeenpartners.nl/ine-poppe

JAN REDZISZ (PL) is a cultural analyst and project manager. His main fields of expertise are CSR and Futurism. He also combines the two areas into a unique service of his own design aimed at boosting an organisation's preparedness for a new technological reality. His toolbox includes: storytelling techniques, foresight games and workshops structured around self-leadership and self-management skills. He provides ideation sessions, lectures on ethical design, curates debate panels, provides editorial services for reports, as well as general in-house consulting and research. His capacity for cross-sector consulting and research comes together in his incredibly humorous short fiction stories.
www.linkedin.com/in/jan-redzisz-26603746/

TOBIAS REVELL (UK) is an artist and designer. Spanning different disciplines and media his work addresses the urgent need for critical engagement with material reality through design, art and technology. Recent work has looked at the idea of technology as a territory, expectations for the future, rendering software as well as the occult and supernatural in pop culture discussions of technology. He is Course Leader of the MA Interaction Design Communication at the London College of Communication, UAL. He is a co-founder of research consultancy Strange Telemetry and one-half of research and curatorial project Haunted Machines who curated Impakt festival in 2017.
www.tobiasrevell.com

MOHAMMAD SALEMY (IR) is an artist, curator and critic based in NYC and Vancouver. He organised the '*Incredible Machines*' conference in Vancouver, 2014. For E-flux he curated *Supercommunity. Art after Machines* in 2015. He has written for Dis Magazine and took part in the *Reinventing Horizons* Symposium (2016). He currently co-organises The New Centre for Research and Practice and - together with researchers from The New Centre - he curates '*Artificial Cinema*', which will also be a part of the ROBOT LOVE Exhibition. This cinema project strives for more open and complex collaborations between humans and machines while exploring the history and future of science-fiction cinema.
www.supercommunity.e-flux.com/authors/mohammad-salemy

EMILIO VAVARELLA (IT) is an interdisciplinary artist and researcher. He graduated summa cum laude from both the University of Bologna with a BA in Visual, Cultural and Media Studies, and from Iuav University of Venice with an MA in Visual Arts. He also studied abroad thanks to fellowships at the Bezalel Academy of Arts and Design, Tel Aviv and Istanbul Bilgi University. Emilio's artistic work has recently been shown at: ISEA, EYE- BEAM, SIGGRAPH, GLITCH Festival, European Media Art Festival, Media Art Biennale and Japan Media Arts Festival. His work has been published in: ARTFORUM, Flash Art, Leonardo and WIRED. He currently lives and works in New York.
www.emiliovavarella.com

INE GEVERS (NL) is a curator, writer and activist. She has been the artistic director of the Niet Normaal foundation which aims to organise art exhibitions and artistic campaigns concerning socially relevant themes for wide audiences since 2007: 'Niet Normaal' (Beurs van Berlage, Amsterdam, 2009), 'Yes Naturally' (Gemeentemuseum, The Hague, 2013) and 'Hacking Habitat: Art of Control' (2016). She is the editor and author of various publications, such as: *Place, Position, Presentation, Public* (Maastricht: Jan van Eyck Akademie, 1992); *Beyond Ethics and Aesthetics* (Nijmegen: SUN, 1996); '*Images that demand Consummation*', in: *Now* (Documentary, 2005); *Niet Normaal: Difference on Display* (Rotterdam: Nai010, 2009); *Yes Naturally* (Rotterdam: Nai010, 2013); *Hacking Habitat* (publication Utrecht: Niet Normaal foundation, distribution Rotterdam: Nai010). **www.nietnormaal.nl / www.inegevers.net**

DENNIS KERCKHOFFS (NL) is a tutor at the interdisciplinary bachelor programme Liberal Arts & Sciences at Utrecht University, his alma mater where he studied molecular biology and gender studies with a focus on feminist critique and technoscience. Dennis has been involved with Niet Normaal Foundation since 2011. For '*Yes Naturally*' (2013) he was both an editor and programme maker. For the next event '*Hacking Habitat*' (2016) he acted as an editor again, as he did for the current project ROBOT LOVE. He has further developed his affection for holistic practices outside academia such as yoga and permaculture during his many travels abroad. In 2015 he started studying Japanese Shiatsu Therapy. **www.uu.nl/staff/DGSKerckhoffs/0**

KLAAS KUITENBROUWER (NL) is a critical media theorist who works for Het Nieuwe Instituut. He has a background in digital culture and has focused on bots and algorithmic culture for the past 18 months. He studies phenomena like the cultural effects of datafication and platformisation (organisations shaping themselves according to the logic of digital platforms), and the (imagined) roles of AI and machine learning systems in society. He has a special interest in the worldview of machines.
www.linkedin.com/in/klaas-kuitenbrouwer-97715a4

LUIS LOBO-GUERRERO (NL) is a Professor of History and Theory of International Relations at the University of Groningen. He is the author of Insuring Security: Biopolitics, Security and Risk, and Insuring War: Sovereignty, Security and Risk, and Insuring Life: Value, Security and Risk. He is director of the Centre for International Relations Research (CIRR), member of the Executive Board of the Groningen Research Institute for the Study of Culture, member of the editorial/advisory boards of the journals Security Dialogue; Journal of Intervention and State Building; and Resilience: International Policies, Practices and Discourses. He is also member of the international advisory board of Revista Pleyade (Chile)
www.rug.nl/staff/l.e.lobo-guerrero

MARTIJNTJE SMITS (NL) is a philosopher specialising in technology. Her most recent academic position was as a Philosophy of Science Professor at Utrecht University. She has also worked for the Rathenau Institute, studying social robotics. She is currently working on a new book 'Frankenstein' for the Frankenstein Year, which is due out in September 2018. The book will be an update of her book 'Monsterbezwering' (2002). Robots are one of the three case studies featured in the book, which also deals with machine technology and human enhancement. She also works with politicians on their lack of attention for technology in their work. She has also lectured on robot love in recent years, but she is still looking for the right perspective to write about this subject.
www.martijntje.nl

IRIS VAN DER TUIN (NL) is a professor of Theory of Cultural Inquiry at Utrecht University (Department of Philosophy and Religious Studies) and programme director of the interdisciplinary bachelor programme Liberal Arts and Sciences (LAS) at Utrecht University. Her background is in gender studies and feminism, which she has written several books on. Her research is situated at the intersection of the philosophy of science and science & technology studies, cultural theory and the new humanities. One of her projects centres on new materialism. She works with students and colleagues on all kinds of issues, some of which also have a special interest in algorithms and combining different fields of inquiry, such as a group on the ethics of coding.
www.uu.nl/staff/IvanderTuin/0

231

ARTISTS

ADVANCED
TELECOMMUNICATIONS
RESEARCH INSTITUTE (ATR),
pp. 36,37, 208, 209
Osaka, Japan
www.atr.jp

JOHANN ARENS
pp. 184, 185
1981, Aachen, Germany
www.johannarens.com

ARMY OF LOVE
ALEXA KAROLINSKI & INGO
NIERMANN
pp. 60, 61
www.thearmyoflove.net

KORAKRIT ARUNANONDCHAI
pp. 170, 171, 218, 219
1986, Bangkok, Thailand
www.c-l-e-a-r-i-n-g.com

ADAM BASANTA
pp. 76, 77
1985, Tel Aviv, Israel
www.adambasanta.com

WILL BENEDICT
pp. 130, 131
1978, Los Angeles,
United States
www.balicehertling.com

MARGRIET VAN BREEVOORT
pp. 95, 196, 208, 209
1990, Amsterdam,
the Netherlands
www.margrietvanbreevoort.nl

BUREAU D'ETUDES
pp. 142, 143
Paris, France
www.bureaudetudes.org

FELIX BURGER
pp. 122, 123
1982, Munich, Germany
www.felix-burger.de

LOUIS-PHILLIPE DEMERS /
BILL VORN
pp. 206, 207
1959, Montreal,
Quebec, Canada
www.processing-plant.com
www./billvorn.concordia.ca

COLOPHON

ROBOT LOVE is an international Expo Experience at the crossroad of art, design & technology. During a 10-week period in the autumn of 2018 Robot Love offers three main programme components to the public: a large-scale exhibition with works by 50 international artists, a publication, a public programme including the All-Inclusive Cyborg Catwalk, and the robot café. **ROBOT LOVE** is the fourth large-scale exhibition organised by the **Niet Normaal Foundation**. The first public outreach of **ROBOT LOVE** was in October 2017 with the Robot Love Embassy during the Dutch Design Week / World Design Event and was organised in co-creation with The Army of Love as a prelude to the 2018 Expo Experience.

ROBOT LOVE EXPO EXPERIENCE
Artistic Director, concept owner and curator
Ine Gevers
Managing Director
Eefje Op den Buysch
Artistic programme
Monique Verhulst, Laura Mudde, Brenda Fischer-Campbell, Joannette van der Veer
Production
Loes Hermans, Mark van Veen, Janneke Koolen, Sanne Heesbeen, Laurence Bilger, Marijke van Ham, Wisse Ruyter , Monique Verhulst
Marketing and Communication
Daniel Bouw, Marieke Verkoelen, Els van Rossum, Snir Gedasi, Vicky Bosch, Chloé Martens, Luc van Acht, Lieve Op den Buysch
Robot Love Academy (education)
Peter de Rooden, Eva Vesseur, Laura Bertram
Robot Love Embassy 2017 at Dutch Design Foundation/ World Design Event
Architecture: Bruno Vermeersch
Artists: Army of Love, Polina Baikina, Jan de Coster,

Edwin Dertien, Gael Langevin, Manuel Pellegrini, Nicole Perez, Tobias Revell, Mirjana Smolic, Paul Segers
Production: Peter de Rooden, Sandra Bosch, Veerle Pennings, Vincent Hoenderop, Krista Janssen
All Inclusive Cyborg Catwalk Winners
Bartosz Seifert, Gill Baldwin, Adrianus Kundert
Cyborg Catwalk Jury
Camille Baker, Charlotte Bik, Marco Donnaramma, Maison the Faux, Martijn Paulen, Anouk Wipprecht
Photography
Peter Cox, Theo Janssen
Graphic Design
Autobahn
Location
Bouwfonds Property Development (BPD) - Campina Milk Factory, Eindhoven
Audiovisual
AV-Registered
Belichting
Hoevenaars
Robotics
RobotXperience, Smart Robot Solutions
Location
Bouwfonds Property Development (BPD) - Campina Milk Factory, Eindhoven
Board
Emile Aarts, Frens Frijns, Anastasia van Gennip, Zuzanna Skalska, Julienne Straatman, Margreth Verhulst
Recommendation Committee
Ute Meta Bauer, Rosi Braidotti, Hendrik Driessen, Vanessa Evers, Vincent Icke, Maarten Steinbuch, Martijn Sanders, Peter Weibel
Advisory Committee
Thom Aussems, Edwin Dertien, Paul Domela, Tonny Groen, Caroline Nevejan, Tim Vermeulen
With thanks to artists and lenders:
Van Abbemuseum, Eindhoven; Andrew Kreps Gallery, New York; Annet Gelink Gallery, Amsterdam; Boros Foundation, Berlin; Büyük Efes Art Collection, Izmir; Centraal Museum, Utrecht, CLEARING New York/

PARTNERS

Funds

BRABANT C

GIESKES·STRIJBIS
FONDS

 Bank Giro Loterij | FONDS

 bkkc brabants kennis centrum kunst en cultuur

BLOCK BUSTER FONDS {

 cultuur eindhoven

FONDS 21

 M mondriaan fund

stimulerings fonds creatieve industrie

  PRINS BERNHARD CULTUURFONDS

VSBfonds, iedereen doet mee

Partners

 AutomotiveNL

 bpd

 BALTAN LABORATORIES

 CultuurStation

  DIMENCO

dutch design foundation

EFFENAAR

 EHV 365

FILOSOFIE MAGAZINE

 FNV

 GLOW

 idfa

nederlands kamerkoor

PAKHUIS DE ZWIJGER*

shapeways*

 SIOUX SOURCE OF YOUR TECHNOLOGY

smartrobot.solutions

 Had van Morgen

SUMMA Zorg

Tilburg University

SINT TRUDO

 TU/e Technische Universiteit Eindhoven University of Technology

UNIVERSITY OF TWENTE.

VAN ABBE MUSEUM

VERBEKE FOUNDATION

VICE

 WageIndicator.org

Special thanks to

 Aim At Art

  arminius congres- en debatcentrum

 THE ART OF ROBOTICS

 ArtechLAB* Amsterdam

 ArtEZ Lectoraat kunst- en cultuureducatie

 Partner BRAINPORT EINDHOVEN

 TUDelft

 d t w 2018

 Fontys SCHOOL OF ENGINEERING

  EINDHOVEN

 HKU

 KUNSTHALLE

 LABKOO®

 Maker Faire Eindhoven

 M MUZIEKGEBOUW FRITS PHILIPS EINDHOVEN

 [night]<of><the> NERDS <5>[juni]

OS △ OOS

 EINDHOVEN PARK THEATER

 SOLID SYSTEMS SECURITY GROUP 24/7

 SiNTLUCAS CREATING OPPORTUNITIES

 WW WEEKEND VAN DE WETENSCHAP

 vpro medialab

240

Brussels; Collectie Joep van Lieshout, Rotterdam;
GALLERIAPIÙ, Bologna; Kraupa-Tuskany Zeidler, Berlin;
Museum Arnhem; Tanya Leighton, Berlin, and those
who wish to remain anonymous.

Special thanks to: Nick Aikens, Hester Alberdingk Thijm,
Joy Arpots, Danielle Arets, Lucky Belder, Jan de Coster,
Maxim Februari, Charles Esche, Annie Fletcher, Johan
Grimonprez, Ton van Gool, Tonny Groen, Giep Hagoort,
Marleen Hartjens, Pieter Jonker, Randal Kerstjens, Henk
Kiela, Johan Kolsteeg, Frank Kresin, Raymond Kuypers,
Gerard Meulensteen, Hilde Meijs, Margot Neggers, Hans
Nijssen, Randall van Poelvoorde, Jan Popma, Jesse
Scholtes, Johanna Weggelaar.

ROBOT LOVE PUBLICATION

Concept & Development
Ine Gevers

Editorial Board
Ine Gevers, Dennis Kerckhoffs, Luis Lobo-Guerrero,
Klaas Kuitenbrouwer, Martijntje Smits, Iris van der Tuin

Image editing
Ine Gevers, Laura Mudde, Vicky Bosch

Text editing English
Brenda Fischer-Campbell, Titus Verheijen

Translation from Dutch
Sam Garrett (Arnon Grunberg)

Proofreading
Titus Verheijen, Theo Janssen

Photography
Peter Arno Broer, Theo Janssen

Chatbot PIP
Ine Poppe, Bart Roorda, Marc Buma, Peter Schilleman

Design
Autobahn

Project Coordination
Wisse Ruyter, Vicky Bosch

Lithography
DPS online, Amsterdam

Typeface
Quicksand & Robot Love (custom made by Autobahn)

Publisher
© 2018 Uitgeverij TERRA
Terra is part of Uitgeverij TerraLannoo bv
P.O. Box 97, 3990 DB Houten, The Netherlands
info@terralannoo.nl / www.terra-publishing.com

First print, 2018
ISBN 978 90 8989 776 3
NUR 740, 656

Niet Normaal Foundation organises visual art
exhibitions with the ambition to initiate public
awareness, debate and generate deliberate human
feedback. Previous editions: *I + the Other/ Ik + de
Ander, Art and the Human Condition* (Amsterdam, 1994),
Difference on Display/ Niet Normaal (Amsterdam,
2009/2010), *Yes Naturally/ Ja Natuurlijk, How art saves
the World* (Den Haag, 2013), *Hacking Habitat, Art of
Control* (Utrecht, 2016), *Robot Love, Can we learn from
robots about love?* (Eindhoven, 2018)
Niet Normaal Foundation holds the ANBI status. Further
information on ROBOT LOVE and earlier editions

www.nietnormaal.nl
www.robtlove.nl / www.robotlove.eu